STELLINGEN

I

Het is in strijd met artikel 5 van het Verdrag van Chicago, om het recht tot het uitvoeren van ongeregelde vluchten afhankelijk te stellen van een vergunning.

II

Het is gewenst, dat artikel 20 van de Wet op de Krijgstucht, mede gelet op artikel 22 van die Wet, wordt gewijzigd in dien zin, dat de strafoplegger beslist of een met plaatsing in een tuchtklasse gestrafte militair in arrest moet worden gesteld.

III

Vervoersongevallenstatistieken, welke betrekking hebben op het aantal gedode passagiers per passagierkilometer, geven maatschappelijk bezien een onjuist en voor wat betreft de luchtvaart een te ongunstig beeld van de relatieve veiligheid van verschillende takken van transport.

IV

Het is gewenst, dat de bevoegdheden van de Raad voor de Luchtvaart worden uitgebreid tot het nemen van maatregelen ten aanzien van enkele nader te bepalen categorien grondpersoneel.

V

De interpretatie van de Hoge Raad van artikel 27 van het Wegenverkeersreglement, houdende, dat een weggebruiker bij het naderen van voorrangsverkeer verplicht is, het *gehele* weggedeelte van een wegkruising vrij te laten, dat gerekend kan worden tot elk der kruisende wegen te behoren, is niet in overeenstemming met de eisen van de practijk van het moderne wegverkeer.

VI

Hefschroefvliegtuigen dienen in verband met hun bijzondere eigenschappen aan een van andere luchtvaartuigen afwijkend regime van luchtvaartvoorschriften te worden onderworpen.

VII

Het verdient aanbeveling de Vreemdelingenwet van 1849 te herzien, daar onder het vigerende stelsel de rechtszekerheid van in Nederland vertoevende vreemdelingen op onvoldoende wijze wordt gewaarborgd.

VIII

De snelle berechting door de politierechter van diefstallen en verduisteringen van goederen bestemd voor of toebehorende aan slachtoffers van de jongste overstromingen is wegens de snelle werking te verkiezen boven een berechting door de gewone rechter, met eventuele langere straffen.

THE AIRCRAFT COMMANDER
IN COMMERCIAL
AIR TRANSPORTATION

THE AIRCRAFT COMMANDER IN COMMERCIAL AIR TRANSPORTATION

PROEFSCHRIFT
TER VERKRIJGING VAN DE GRAAD VAN
DOCTOR IN DE RECHTSGELEERDHEID
AAN DE RIJKSUNIVERSITEIT TE LEIDEN,
OP GEZAG VAN DE RECTOR MAGNIFICUS
Dr J. J. L. DUYVENDAK, HOOGLERAAR IN
DE FACULTEIT DER LETTEREN EN WIJS-
BEGEERTE, TEGEN DE BEDENKINGEN
VAN DE FACULTEIT DER RECHTSGE-
LEERDHEID TE VERDEDIGEN OP
WOENSDAG 1 APRIL 1953 TE 16 UUR

DOOR

MENNO SJOERD KAMMINGA

GEBOREN TE TANAH BESI,
NEDERLANDS INDIË (THANS INDONESIË)
IN 1916

'S-GRAVENHAGE
MARTINUS NIJHOFF
1953

Promotor: Prof. Mr D. GOEDHUIS

ISBN 978-94-011-8671-1   ISBN 978-94-011-9467-9 (eBook)
DOI 10.1007/978-94-011-9467-9

" .... the future development of international civil aviation can greatly help to create and preserve friendship and understanding among the nations and peoples of the world ...."

Preamble to the Chicago Convention.

CONTENTS

PART II. THE DRAFT CONVENTION ON THE LEGAL
          STATUS OF THE AIRCRAFT COMMANDER

INTRODUCTION

Less than fifty years have elapsed since the first flight of a power-driven aircraft. On December 17, 1903, the Wright brothers made their now historic flights in an aircraft which they had designed and built themselves; at the fourth attempt on that day a distance of 852 feet was covered during a flight which lasted 59 seconds [1].

Unparalleled technical advances followed those first hesitant steps, with the result that aviation has developed into a factor influencing practically every field of society [2].

The complex mechanism of modern commercial aviation can only function through the combined efforts of countless people. The inspiration of aircraft constructors, the insight and perseverance of air pioneers and the conscientious work of the ground personnel all play their part.

Among those who share in this joint task, however, the aircraft commander occupies a special place. He finds himself at the head of a small but comparatively isolated community, which may come under different jurisdictions within a short space of time. From a purely academic aspect he is therefore an interesting figure. Moreover, it may also be useful to examine his legal status for more practical reasons. After all, the aircraft commander holds a key position in aviation, as the safety, economy and regularity of the flight often depend on his skill and judgment.

The starting-point of the following study is the aircraft commander's legal status in positive law; as we shall come across various deficiencies, an attempt will subsequently be made to indicate how his legal status might be amplified and regulated in further detail.

Two limitations are observed in this connection.

---

[1] Kelly, "Miracle at Kitty Hawk" page 114.
[2] Cf. Ogburn, "The Social Effects of Aviation"; Wright, "Aviation's Place in Civilization", R. Ae. S. Aeronautical Reprint No. 101.

1

Firstly, our observations will only relate to the aircraft commander in commercial air transportation.

It is true that in principle private pilots and commercial pilots may find themselves in a position similar to that of the airline pilot, and in general they are subject to the same rules; but for practical considerations it is quite reasonable to make a distinction between these categories.

Secondly, an assessment of the status of the aircraft commander in the social structure, or in other words, an inquiry into the level of his profession, does not come within the scope of our study. It may be mentioned in passing, however, that opinions on this point vary from the idea that pilots are professional daredevils, "professionnels du risque, qui ont véritablement fait du danger leur métier",[1] to the view that "because of his responsibilities for human life the pilot is in the same professional category as a doctor."[2]

THE AIRCRAFT COMMANDER IN POSITIVE LAW

Whatever the role of the aircraft commander in commercial air transportation may be, on the whole his legal status in positive air law has largely been neglected.

Unlike the master of a ship, whose legal status is based on both codified law and tradition, the commander of an aircraft lacks a clearly defined status.

Although numerous national legislations contain provisions relative to the aircraft commander, these are characterized by a great diversity in structure and contents rather than a complete and exhaustive treatment of the subject[3].

A few legislations give detailed rules and regulations but they are mostly derived from maritime law. In general, however, the aircraft commander is merely assigned one or two summary attributes, or else his status is not indicated in any way whatsoever.

---

[1] Laurent Eynac, cited by Le Goff, "La Loi du 25 Mars 1936", Droit Aérien 1936, page 146.

[2] W. A. Patterson, President of United Airlines, Aviation Week, May 29th, 1950, page 44.

[3] Compilations of legislations concerning the aircraft commander are to be found in Maschino, ,,La Condition Juridique du Personnel Aérien"; Sandiford, ,,Lo Stato Giuridico del Commandante di Aeromobile"; ICAO Doc. 4538/LC 18.

In view of the diversity of the national rules and regulations, not to mention their obvious shortcomings and defects, we have decided not to undertake a detailed comparative study of the existing legal provisions.

It must be borne in mind that the usefulness of *national* legislations — no matter how perfect they may be in themselves — is limited in this respect. It is of little benefit to the commander, that his national law should give him certain powers, if he cannot avail himself of these rights while under another jurisdiction. If regulations governing the legal status of the aircraft commander are to have any practical value, they must be *internationally* effective. At present, however, comprehensive international regulations of this nature are non-existent.

Current Dutch air law has been selected as our starting-point for the following observations. Not so much because the Dutch legislation might serve to illustrate how the legal status of the aircraft commander ought to be regulated, but because it provides a good example of the difficulties and uncertainties with which the aircraft commander may be faced.

Many of the Dutch regulations are derived from or based on international conventions, so that the problems mentioned are universal. In cases where this does not apply, or where Dutch law is completely silent, we have indicated the solutions given in maritime law or in the air legislation of other countries.

THE PLAN

This study consists of two parts.

The first part deals with the aircraft commander's position under Dutch law.

At first it seemed to be a good idea to follow the system which Cleveringa adopted in his ,,Het Nieuwe Zeerecht'' when discussing the position of the captain of a sea-going vessel. Although this arrangement is logical — the captain as master of the ship, as representative of the owner, and as employee — some difficulties arose in applying it to the aircraft commander.

The above-mentioned arrangement of the subject matter is roughly the same as that of Articles 341–392 in the third title of the second book of the Netherlands Commercial Code (the section concerning the master of a ship). As far as the aircraft commander is concerned, however, a systematic guide of this nature is non-existent. If it is possible to speak of a settlement of his legal status at all, this is chiefly to be found in scattered bits and pieces. A second objection to adopting the system outlined is that it would involve taking up a definite standpoint in regard to a far-reaching analogy between the master of a ship and the commander of an aircraft.

An alternative was to deal separately with the rights and obligations of the aircraft commander. Such an arrangement likewise proved to be somewhat unsuitable for a discussion of the codified law, since — as will be seen later — the rights are conspicuous by their almost total absence.

The next arrangement considered was one which appears in many books on air law, viz. division into international and national public and private law. Apart from the fact that there are objections to such a division in itself (the Brussels Convention on assistance and salvage, for example, contains elements of both private and public law), the results of such a scheme proved to be confusing mainly through the partial duplication of international and national air law. National air law is largely derived from international air law, though sometimes delay occurs.

The arrangement finally selected for the first part is as follows.

To begin with, the statutory legislation applicable to the aircraft commander is roughly outlined in a chapter entitled "The Authorities and the Aircraft Commander".

This is followed by three chapters dealing with "Transport Aircraft", "The Crew" and "Flight Operations", respectively, and subsequent chapters on "Instructions from the Ground" and "Search and Rescue". Next come two chapters entitled "Sanctions" and "Liability", dealing with the criminal proceedings, civil proceedings and disciplinary measures which may be taken against the aircraft commander. The final chapter discusses "The Aircraft Commander as Employee".

As will be seen from the first part, the regulations concerning the legal status of the aircraft commander have a number of shortcomings. The second part of our study is therefore devoted to a draft convention regulating the legal status of the aircraft commander, with the object of ascertaining the extent to which it meets the requirements. This part is subdivided into a historical review and a discussion of each article, and it concludes with some proposals to amplify the draft.

SOURCES

A number of manuals of air law must first be mentioned. Although these contain observations on legal problems connected with our subject, they only deal summarily with the legal status of the aircaft commander.

LEMOINE goes into this question in more detail, however, on the basis of the relevant French legislation.

A distinction must be made between the works published prior to the Chicago Convention (GOEDHUIS, LE GOFF, KROELL, LUPTON, McNAIR, ZOLLMANN) and those published afterwards, viz. RIESE, LEMOINE, SHAWCROSS and BEAUMONT (2nd edition), and FIXEL (3rd edition).

The most recent manuals consulted were those of CHAUVEAU (1951) and DE JUGLART (1952). CLEVERINGA's manual on maritime law contains a convenient analysis of the legal status of the ship's master.

For guidance on the organizational and technical background of commercial aviation, the works of the following American authors were consulted: FREDERICK, PUFFER, WILSON and BRYAN, NICHOLSON, WOLFE, SPEAS, BAKER.

Literature about flying personnel in general, or the aircraft commander in particular, is comparatively scarce besides being frequently out-of-date. A first article on the subject of "La situation juridique des aéronautes en droit international" was written by WILHELM in 1891, i.e. twelve years before the first flights of the Wright brothers.

Monographs on the status of flying personnel have been written by MASCHINO (thesis — Paris, 1930), GIANNINI (1937), BUCHER (thesis — Lausanne, 1949) and BRATSCHI (thesis — Berne, 1951), while SAVOIA (1929) and SANDIFORD (1934) have published treatises on the aircraft commander.

A number of articles have also appeared in the course of the years; their authors, listed in chronological order, are as follows: WÜSTENDORFER (1931), RIESE (1932), RICHTER (1932 and 1935),

BABINSKI (1932), LE GOFF (1936), DÖRING (1937 and 1941), CHARLIER (1947) and KNAUTH (1947). The last two articles are particularly important because they take into account the practices adopted in post-war aviation.

In the Netherlands, GOEDHUIS (1933) and HONIG (1951) have written about this subject.

The official records and reports of CITEJA, and more recently the documents of ICAO, naturally provide a source of information, and the same applies to some IATA Bulletins. These publications contain the reports drawn up by THIEFFRY, BABINSKI, GARNAULT and BEAUMONT.

In order to obtain a better insight into the duties and everyday problems of the aircraft commander, it proved to be indispensable to consult a number of aviation journals of a technical and general nature.

It must also be mentioned that in the case of a study such as the present one, where the subject is highly topical, contact with aviation in actual practice is almost essential.

Insofar as certain conclusions could not be arrived at from personal observation, conversations with numerous people employed in aviation often threw light on the matter.

PART I

THE AIRCRAFT COMMANDER IN DUTCH LAW

Chapter I

THE AUTHORITIES AND THE AIRCRAFT COMMANDER

Many different considerations have led the authorities to control aviation on military and economic grounds as well as for safety reasons. The activities of even the earliest aeronauts immediately led to the issue of government edicts and decrees [1].

We shall not attempt to answer the question of whether the development of aviation was excessively restricted and hampered by this. It may be useful, however, to begin our study with a brief account of the structure of the system of statutory rules and regulations under which the aircraft commander performs his daily duties.

Aviation is characterized by two special features: it is of an international nature and it is extremely susceptible to technical progress. Both of these characteristics have had their effects on the system at present in force.

Our remarks will be confined to the broad outlines of the system and therefore relate solely to the Chicago Convention and its Annexes as well as the Netherlands Aviation Act and the orders issued under that Act. We shall disregard other sources of public law such as the Paris Convention (1919), which is merely of historic interest nowadays, and the International Sanitary Convention concluded at The Hague in 1933 (together with the resultant Dutch legislation), as being of less fundamental importance to the aircraft commander. A separate paragraph [2] will be devoted to the Netherlands Air Accident Act.

---

[1] Order of the French Police of 23rd April 1784, controlling the ascent of "Montgolfières" and order of the Hamburg municipal authorities of 18th August, 1786, cited by Riese, "Luftrecht", page 10. For a description of these first flights see Davy, "Interpretive History of Flight"; Fuld, "Van Icarus tot Zeppelin".

[2] See page 80.

THE CHICAGO CONVENTION

Towards the end of World War II, when it became clear that international air traffic would assume undreamt-of proportions after the war, the United States of America took the initiative in calling a conference to regulate international civil aviation [1]. For this purpose the U.S.A. extended an invitation to each of the Allies and a number of neutral States. The invitations were accepted by all those invited, except Saudi Arabia and Russia. The former did not acknowledge the invitation at all, while Russia raised objections to sitting at the conference table along with Spain, Portugal and Switzerland. It is questionable whether this was the real reason for refusal, however. Russia may have wished to reject any international control of aviation over its territory as a matter of principle [2].

Delegates of 54 countries subsequently met at Chicago during the months of November and December, 1944.

Of the agreements concluded on that occasion, the Convention on International Civil Aviation (hereafter referred to as the "Chicago Convention") is particularly important in connection with our subject.

The Chicago Convention, which superseded the Paris Convention of 1919 and the Havana Convention of 1928 [3], has now been ratified by 57 States [4]. In the Netherlands the Chicago Convention was ratified by the Act of February 28, 1947 and it came into force on April 25, 1947.

Both the original text and a Dutch translation were published by inclusion in the Netherlands Statute Book (Decree of June 3, 1947).

It is assumed that the rules of law contained in the Convention immediately became applicable to Dutch citizens through the ratification and publication [5].

---

[1] Riese, "Luftrecht" page 97; Pépin, "Le Droit Aérien", Recueil des Cours de l'Academie de Droit International 1947, page 490.

[2] Wagner, "Les Libertés de l'Air" page 91.

[3] Article 80 of the Chicago Convention.

[4] In the meantime Guatemala has denounced the Convention, which denunciation will be effective on 1st June, 1953.

[5] Cf. Goedhuis, "Handboek voor het Luchtrecht" page 323; Telders, "Le droit des gens dans la jurisprudence des Pays-Bas" page 3, and the jurisprudence mentioned therein; Langemeyer, "Inleiding tot de Studie van het Nederlandsche Recht" page 84; to the same effect Riese, "Luftrecht" page 61.

The aim of this Convention, as envisaged by the contracting States, cannot be better expressed than by quoting the following extract from the preamble: " ... the undersigned governments having agreed on certain principles and arrangements in order that international civil aviation may be developed in a safe and orderly manner and that international air transport services may be established on the basis of equality of opportunity and operated soundly and economically, have accordingly concluded this Convention to that end".

To what extent has the Convention fulfilled this lofty purpose? There is reason to doubt whether the establishment of international air transport services on a "basis of equality of opportunity" and operation "soundly and economically" have been generally achieved [1].

In regard to the development of international civil aviation "in a safe and orderly manner", however, good results have been secured, mainly due to a number of technical regulations contained in the Annexes to the Convention, which will be dealt with in further detail on the following pages. It is important to note that through the participation of the United States the Chicago Convention has acquired a more universal character than was the case with the Paris Convention. [2]

The Chicago Convention is divided into four parts, entitled:

Air Navigation, The International Civil Aviation Organization, International Air Transport and Final Provisions.

The principles and regulations contained in Part I, Air Navigation, are of particular interest to the aircraft commander, since this Part gives a large number of statutory regulations concerning such matters as prohibited areas, customs airports, rules of the air, entry and clearance, prevention of spread of disease, registration and display of marks, facilitation of formalities, customs and immigration procedures, aircraft in distress, investigation of accidents, documents to be carried, radio equipment, certificates of airworthiness, certificates of competency, log books and dangerous cargo.

We shall not discuss the material contents of these regulations

---

[1] See Riese, "Luftrecht" page 107 onwards and page 129; for a more extensive version see Wagner, "Les Libertés de l'Air".
[2] Cf. Goedhuis, "Air Law in the Making" page 27.

at this point. Insofar as they are of importance to the captain, they will come up for discussion later.

THE ANNEXES TO THE CHICAGO CONVENTION

The detailed technical regulations embodied in the Annexes to the Chicago Convention are exceptionally important for the aircraft commander in the exercise of his profession. When one considers that there is perhaps no other form of human activity endowed with such an international character as civil aviation, it is obvious that the existence of internationally accepted and applied standards and procedures is a first essential for safe and efficient air navigation [1].

Within the space of a few hours one can fly across several national frontiers, and it would be a dangerous and untenable state of affairs if the aircraft commander repeatedly had to cope with widely varying instructions, procedures and situations in the course of a flight. By means of the "Standards and Recommended Practices" contained in the Annexes to the Chicago Convention, an attempt has been made to secure uniformity in many different fields connected with the execution of a flight. To a certain extent the aircraft commander can be sure that meteorological reports are drawn up in the same code in all the countries which are members of the International Civil Aviation Organization (ICAO); that the charts to be used by him satisfy certain minimum requirements; that the dimensions and construction of airfields, as well as the airfield equipment, fulfil minimum specifications which are also known to him; that the instructions from the control towers will be given in a standard manner; that his colleagues in the air are observing the same traffic rules and have a certain minimum degree of experience; that aircraft and components, no matter where they are manufactured, guarantee a minimum safety level.

Before ascertaining how far the Annexes serve their purpose, it may be appropriate to give a brief account of the organization of ICAO and to explain how an Annex comes into being [2].

---

[1] Cf. Le Goff, "The Present State of Air Law" page 2; Nicholson, "Air Transportation Management" page 84.

[2] Rules of Procedures for and Directives to Divisions, ICAO Doc. 5417-AN/626; Relative Functions of the Council, the Air Navigation Commission, the Divisions and the Secretariat in the Development of Annexes to the Convention, Doc AN-WP/512.

The organs of ICAO include amongst others the Assembly, the Council and the Air Navigation Commission.

In principle the Assembly meets annually; a full scale Assembly is held once every three years, and the other Assemblies only deal with budgetary and administrative matters. All contracting States have an equal right to be represented at the meetings of the Assembly and they are each entitled to one vote (Art. 48).

The Council is a permanent body responsible to the Assembly. It is composed of twenty-one contracting States elected by the Assembly for a period of three years (Art. 50) [1]. Holland is among the countries represented on the Council. The functions of the Council, as listed in Art. 54, include the following: "The Council shall ... Adopt, in accordance with the provisions of Chapter VI of this Convention, international standards and recommended practices; for convenience, designate them as Annexes to this Convention; and notify all contracting States of the action taken; Consider recommendations of the Air Navigation Commission for amendment of the Annexes and take action in accordance with the provisions of Chapter XX".

The Air Navigation Commission is composed of twelve members appointed by the Council from among persons nominated by the contracting States. These persons must have suitable qualifications and experience in the science and practice of aeronautics (Art. 56).

The Commission must consider, and recommend to the Council for adoption, modifications of the Annexes. In addition, it must establish technical subcommissions on which any contracting State may be represented if it so desires (Art. 57).

Although the Chicago Convention does not make any mention of the "Technical Divisions" of the Air Navigation Commission, they may really be regarded as subcommissions within the meaning of Art. 57. The Standards and Recommended Practices are actually formulated and the material contents of the Annexes largely decided upon at these meetings of the Technical Divisions. The meetings are normally held every two years and they are attended by technical experts on the subjects covered by the

---

[1] For some time there has been a vacancy in the Council so that there are only twenty members at the moment; the question has therefore arisen whether the Council is "legally" constituted.

Division's terms of reference. It is important to note that these meetings are also attended by observers from international bodies such as IATA [1] and, to a lesser extent IFALPA [2].

This ensures that practical requirements are taken into account as far as possible in drawing up the regulations.

The "International Standards and Recommended Practices" drafted by the Divisions are subsequently circulated to the States for their comments. Both the proposals of the Division and the comments of the States are then discussed by the Air Navigation Commission. If approved — with or without alteration — the proposals are put before the Council. For adoption by the Council a majority of at least two-thirds is necessary, after which the proposals are again submitted to the States. Unless the majority of the States have announced their disapproval within a specified period (normally 120 days) the Annex then becomes effective (Art. 90).

Under the terms of Art. 37 the States are obliged to bring their national regulations or procedures into line with the international standards and recommended practices and procedures adopted by ICAO — as far as possible. Should they consider this to be impracticable in certain respects, the deviations as compared with the international standards must be notified to ICAO (Art. 38).

ICAO has issued the following fourteen Annexes:

Annex 1 Personnel Licensing
Annex 2 Rules of the Air
Annex 3 Meteorological Codes
Annex 4 Aeronautical Charts
Annex 5 Dimensional Units to be used in Air-Ground Communication

---

[1] International Air Transport Association, union of over 60 scheduled airlines, residing in Montreal, its aims being: a) to promote safe, regular and economical air transport, b) to provide means for collaboration among the air transport enterprises engaged directly or indirectly in international air transport service, c) to cooperate with ICAO and other international organizations.

[2] International Federation of Air Line Pilots Associations, comprises 19 associations of transport pilots with a total of about 10,000 associate members. The Federation is established in London and has amongst other things the following purposes: — a) To promote the interests of the air line piloting profession. b) To aid in the establishment of fair rates of compensation, maximum hours of employment and uniform principles of seniority. c) To foster the passage of legislation to improve the safety of working conditions.

Annex 6 Operation of Aircraft. International Commercial Air
          Transport
Annex 7 Aircraft Nationality and Registration Marks
Annex 8 Airworthiness of Aircraft
Annex 9 Facilitation of International Air Transport
Annex 10 Aeronautical Telecommunications
Annex 11 Air Traffic Services
Annex 12 Search and Rescue
Annex 13 Aircraft Accident Inquiry
Annex 14 Aerodromes

These Annexes are constantly in process of amendment and
revision.

How are these Annexes to be judged, and what influence have
they had on the standardization of international air navigation?

On the one hand it has been pointed out — especially by IATA
— that there are objections to excessive regulation. Riese [1] also
draws attention to "die Gefahr eines Überhandnehmens der Büro-
kratie, die, schon um ihre Existenzberechtigung zu beweisen,
versucht sein könnte eine übermässige Reglementierung einzu-
führen, statt der Praxis auf allen Gebieten, die nicht unmittelbar
die Allgemeinheit berühren, möglichst freie Hand zu lassen".

On the other hand, the establishment of these technical regulations may be looked upon as one of the most important results
of ICAO's activities. Le Goff once remarked, speaking of the
Annexes drawn up during the Chicago Conference: "Had the
conference achieved this result only, it would have justified the
time and effort spent [2]". It is undoubtedly correct to assert that
the Annexes "constitute considerable international agreement
painstakingly reached through coordination between Contracting
States by means of technical discussion extending over a considerable period." [3] Although there is no reason to be dissatisfied
with the contents of the Annexes, the position with regard to the
actual incorporation of the standards in the national legislations
is somewhat different.

As already remarked, the Annexes do not have immediate

---

[1] „Luftrecht" page 108.
[2] Le Goff, "The Present State of Air Law" page 16.
[3] Annex 6, page 38.

binding force but the States are bound to take the necessary steps to put them into effect.

This obligation, however, is limited by expressions such as "so far as it may be found practicable", "to the greatest possible extent", "the highest practicable degree of uniformity", etc. [1].

The above system differs from the procedure applicable under the Paris Convention, whereby CINA could draw up standards which were immediately binding for all contracting States [2]. Riese [3] draws attention to the fact that there are serious drawbacks to the ICAO procedure because in this way international unification can only be achieved incompletely and with a great deal of delay. In our opinion Riese's view is correct, and the present situation in the Netherlands with regard to technical regulations on the subject of air navigation may serve to confirm this, as the Dutch "Regulations for State Control of Air Navigation" (R.T.L.) and the "Air Traffic Regulations" (L.V.R.) are still in force even though they contain regulations which are very much out-of-date.

THE NETHERLANDS AVIATION ACT AND THE ORDERS ISSUED UNDER THIS ACT

A bill to control the use of aircraft and airships, the "Flying and Aviation Bill", was introduced in the Lower House of the Dutch Parliament as early as 1911. After a number of additions and amendments had been made in 1919 (partly on account of the Paris Convention which was concluded in that year), the Aviation Act was ultimately promulgated in the Statute Book on July 30, 1926. Some alterations were subsequently enacted on December 12, 1935.

It might well be assumed that the regulations contained in the Aviation Act would mostly be obsolete nowadays, in view of the enormous changes which have taken place in the scope and practice of air navigation since the Act was passed. As the greater part of the Aviation Act largely consist of general principles, however, this is only true to a slight extent.

---

[1] Cf. Lemoine, "Traité de Droit Aérien" page 59.

[2] Cf. le Goff, loc. cit. pages 7 and 14; comparing the CINA and ICAO this author is, however, of the opinion that with the ICAO "the power to issue international regulations remained intact", an opinion we do not share.

[3] "Luftrecht" page 111.

Art. 52 of the Aviation Act mentions a number of subjects on which instructions will be issued "by general administrative order in the interests of public order and safety". This has been done in the Regulations for State Control of Aerial Navigation, which, as previously stated, are by no means up-to-date.

The Order in question, No. S. 454, was made on December 6, 1928, and has since been repeatedly amended. It contains a number of detailed rules and regulations, many of which — like the Aviation Act itself — are still based on the Paris Convention of 1919, and especially the Annexes to that Convention. It is not surprising therefore, that the majority of these technical instructions have hardly kept up with modern aeronautical development.

Mention must also be made of the Air Traffic Regulations (Decree of February 28, 1929, No. S. 67) which, in their present form, are likewise practically useless for modern air navigation.

The Annexes to the Chicago Convention are not incorporated in the R.T.L. or in the Air Traffic Regulations, although Holland is bound to apply the ICAO Annexes and actually does so. Only on a few minor points has Holland given notice of a deviation from an Annex.

The result of this is that the examinations for pilots' licences are conducted in accordance with examination rules which are largely based on the relevant ICAO regulations and *not* on the widely divergent provisions of the R.T.L.; that aircraft inspections are conducted in accordance with ICAO directives and *not* in accordance with the statutory regulations which are still in force; that traffic is controlled in accordance with the "Rules of the Air" laid down by ICAO and *not* in accordance with the Dutch Air Traffic Regulations [1]. Obviously this has given rise to an extremely undesirable situation, which often confronts the aircraft commander with almost unsolvable problems, makes it difficult to impose sanctions and might lead to serious consequences in civil actions.

The Chicago Convention and its Annexes, as well as the Netherlands Aviation Act, the R.T.L. and the Air Traffic Regulations, contain a number of provisions relating to the rights and especially the obligations of the aircraft commander.

---

[1] Cf. Honig, "Overheidsaansprakelijkheid en Luchtvaart" N.J.B. 1951, page 767.

In the following pages we shall study some of these provisions which affect the commander's legal status.

When we refer to ICAO regulations in this connection it must therefore be remembered that officially they are not yet effective in the Netherlands, though they are actually applied and will be given force of law in this country in the near future. As far as the provisions of Dutch legislation are concerned, it must also be borne in mind that perhaps they are no longer actually applied and will soon be altered.

Le Goff has given a strikingly accurate description of the continuing development of air law: "Gradually rules of national legislations will be replaced by international rules. Sovereignty vanishes, internationalism appears. National laws will have the tendency to disappear in international rules. National legislations creating the conflict of laws, are no more the real source of air law, it is in the ICAO at Montreal" [1].

A new Dutch Aviation Act, Regulations for State Control of Air Navigation, and Air Traffic Regulations which will be better adapted to the requirements of modern aviation and the rules of the Chicago Convention are now being drafted.

All in all, however, Dutch public air law is at present in an extremely inconvenient transitional stage.

THE AIRCRAFT COMMANDER AS AN AGENT OF THE AUTHORITIES

Like the master of a ship, from time to time the aircraft commander is called upon by the Authorities to put certain provisions of the law into effect.

This aspect of his legal status — which Cleveringa [2] compares with self-government — is more highly developed in maritime law than in air law. Nevertheless the aircraft commander is also required to perform certain statutory duties.

This applies primarily in regard to his obligations in connection with the flight documents. Art. 29 of the Chicago Convention specifies the documents which must be carried by an aircraft engaged in international navigation. More details — still based on

---

[1] Le Goff, "The Present State of Air Law" page 24.
[2] Cleveringa, ,,Het Nieuwe Zeerecht" page 214.

the practically identical provisions of the Paris Convention [1] —
are to be found in Chapter X of the R.T.L., "Documents to be
carried in an aircraft." [2]

The commander has two duties in this connection. In the first
place, the commander of an aircraft in flight must see that the
following documents are present on board the aircraft: journey
log book, signal log book, [3] certificate of registration, crew's certificates of competency, certificate of airworthiness, authorization
to install and to operate radio telephony and radio telegraphy
equipment (if any) [4] and, in the case of transport aircraft, a passenger manifest, cargo manifest and air consignment notes [5].

With regard to the journey log book, the commander has a
further special task: he is obliged to see to it that the necessary
entries are made in it immediately after the conclusion of each
flight or journey and, in the case of a journey lasting several days,
daily during that journey [6]. Thus the aircraft commander does
not have to make the entries himself but he is personally liable
to a penalty if the regulation is not complied with.

The entries relate to the names of the members of the crew and
their functions on board, the place and the time of commencement and termination of each flight, forced landings, accidents,
repairs and health measures, if any [7]. This is less than has to be
put into a ship's log book, in which "everything of any importance
occurring on the voyage" is carefully entered [8].

In regard to the signal log book, the R.T.L. merely states that
a record must be kept in it of the place, date and time at which
a signal is received or transmitted and the stations with which
messages have been exchanged. The keeping of the signal log
book is the responsibility of the owner of a transport aircraft and
not of the commander [9].

---

[1] Article 19.

[2] This chapter also deals with the aeroplane and engine logbooks, which documents,
however, — contrary to what the title of the chapter suggests — need not be on board.

[3] Article 184 ter.

[4] Article 185.

[5] Article 185 bis.

[6] Article 184 quater.

[7] Article 184 quater.

[8] Article 348, par. 1 W.v.K., cf. Cleveringa, "Het Nieuwe Zeerecht" page 247.

[9] Article 184 sexies and article 184 par. 2. In addition, these regulations will have
to be altered now that radio telephony has taken the place of radio telegraphy in many
cases; registration of all messages is then impracticable, whilst at the same time the

Entries in the journey log book as well as in the signal log book must be made in ink or with an indelible pencil; entries must not be blotted out or erased, nor may pages be removed. Furthermore, the information intended for insertion in the journey log book may first be put in a "notebook" (i.e. not on loose scraps of paper or on the back of an envelope!) and later transferred to the journey log book [1].

The aircraft commander also has some administrative duties to perform when public health measures are applied to prevent the spread of disease, [2] when goods are imported or exported by air, [3] and when complying with the regulations concerning aerial photography [4].

The extent of the captain's administrative duties under maritime law is much wider. In the first place we have his obligations concerning the registration of births and deaths and also the making of wills.

Such an obligation (and power) is not possessed by the aircraft commander. As cases of birth and death may occur on board aircraft — and have already occurred in actual practice [5] — it is desirable to draw up regulations on this subject for aviation as well, in order to prevent legal insecurity. Provision is made for this in the draft convention to regulate the legal status of the aircraft commander [6]. The power to make wills, for the time being at least, seems superfluous as far as the aircraft commander is concerned. The same applies to the power to perform marriages valid in civil law. In our opinion the legislator is not called upon to provide for such odd occurrences as marriages on board aircraft. [7]

---

necessity hereto is reduced now that more and more communication is automatically recorded with wire recorders on the ground.

[1] Article 184 septies.

[2] Act of 26th October, 1935, S. 626 for the Regulation of the Sanitary Controlof Aerial Navitation and Decree of 11th February, 1936, S. 840; for the danger of contagion as a consequence of air navigation see McFarland, "Human Factors in Air Transport Design" pages 165—208; Ogburn, "The Social Effects of Aviation" page 381.

[3] Decree of 27th December, 1928, S. 500 and Resolution of the Minister of Finance of 17th July, 1929, No. 169.

[4] Decree of 26th April, 1940, S. 540.

[5] A first case of childbirth in the air reportedly took place in 1889 on board a balloon, see Bonnefoy, "Le Code de l'Air" page 216. Airline members of IATA do not generally transport women who are presumptively more than seven months pregnant.

[6] See page 150.

[7] Cf. Coquoz, "Les Perspectives d'avenir du Droit Privé International Aérien", R.G.D.A. 1938 page 36; Lemoine, "Traité de Droit Aérien" page 205.

The master of a ship also plays a part in drawing up the ship's protest [1], in social welfare (repatriation of distressed seamen) [2], and in criminal proceedings (transportation of condemned and suspected criminals) [3]. We consider that provisions of this nature are at present unnecessary in commercial aviation.

Another important point is the master's power to act as a police official when crimes are committed at sea [4]. In principle, a regulation of this nature might also be of importance to the aircraft commander, though in our opinion it would first be necessary to have a more general regulation concerning his power over the persons on board an aircraft. In view of the special character of air transport, the granting of police authority to the aircraft commander is not particularly urgent at present; accordingly such powers are not granted in the draft convention on the legal status of the aircraft commander [5].

To sum up it may be said that the aircraft commander as an agent of the Authorities has a number of administrative duties and powers, but that the extent and nature of these obligations and powers are not as great as those of the master of a ship.

---

[1] Cf. Cleveringa, "Het Nieuwe Zeerecht" page 250.
[2] ibidem page 253.
[3] ibidem page 254.
[4] ibidem pages 254 and 230.
[5] See page 164.

CHAPTER II

TRANSPORT AIRCRAFT

AIRCRAFT IN GENERAL

Without aircraft there can be no air navigation; it is therefore
obvious that we must devote some consideration to the vehicle
commanded by the subject of our study, a vehicle in which he
performs an important part of his daily work.

In the first place we must decide what is meant by an aircraft.

Although the word "aircraft" can hardly be misunderstood in
actual practice, it has been found extremely difficult to draw up
a legally acceptable and technically serviceable definition of an
aircraft. [1]

The most familiar definition is the one which appeared in similar form in Annex A of the Paris Convention and which is now
included in Annex 7 of the Chicago Convention, viz.: "Any
machine that can derive support in the air from the reactions of
the air". An objection to this definition is that it is so wide as to
cover children's kites, toy balloons, meteorological balloons,
model aircraft, etc. [2]

Although the wording does not appear to exclude parachutes,
apparently the definition was not intended to include them, for
at least parachutes are not mentioned in the different subsections
of Annex 7. [3]

Various national legislations contain definitions which are more
restrictive than the one quoted above.

Those used in the Netherlands are briefly mentioned below.

The wide definition already quoted has only been adopted in

---

[1] Cf. Riese, "Luftrecht" page 187.
[2] ibidem page 188.
[3] To the same effect Goedhuis with respect to Annex A of the Paris Convention in
"Handboek voor het Luchtrecht" page 59; Riese, however, appears to consider
parachutes as coming under aircraft, see "Luftrecht" page 185.

its original form in the Air Traffic Regulations (Art. 1, III) and in the Netherlands Code of Civil Procedure (Art. 770).

The Aviation Act contains a limitative list of the contrivances to be regarded as aircraft, viz.: "aeroplanes, airships, free balloons, captive balloons and kites" (Art. 1, II).

The definition employed in the Paris and Chicago Conventions appears in the R.T.L., but with the following modification: "excepting captive balloons, kites and gliders" (Art. 1).

In the Act regulating the Sanitary Control of Aerial Navigation, the definition is likewise extended and at the same time restricted through the addition of "and which is intended for aerial navigation" (Art. 1, Par. 1b).

Lastly, it is proposed to include the following definition in the new Aviation Act:

"Aircraft: Machines which can be supported in the atmosphere through forces exerted on them by the air, with the exception of such machines as may be designated by general Order" (Art. 1, Par. 1b).

It is understood that toy balloons, etc., are to be excluded from the scope of the Aviation Act in this manner. Although there is thus a tendency to restrict the meaning of the term "aircraft" a much wider definition applies in the United States, viz.:

"Any contrivance now known or hereafter invented, used, or designed for navigation of or flight in the air" (Civil Aeronautics Act, 3.1.4). Both Shawcross [1] and Riese [2] are of the opinion that this definition also covers rockets and projectiles.

Cooper [3] remarks, however, that the normal terms such as "air law", "air navigation" and "aircraft", if taken in a literal sense, only apply to that part of space where air is found, or in other words, the atmosphpere. If these terms are to be applied to flights outside the atmosphere, it would be advisable to speak of "law of space and flight", "flight" and "flight instrumentality".

Further comparison of the above definitions is unnecessary, since the transport plane with which we are particularly concerned, is covered by them in all cases.

---

[1] Shawcross and Beaumont, "Air Law" page 15.
[2] Riese, "Luftrecht" page 188.
[3] Cooper, "Air Law — A Field for International Thinking", Transport and Communications Review, October-December 1951, page 6.

We shall concentrate our attention on the commander of the transport aeroplane, as airships and free balloons are very seldom used in civil aviation nowadays.

The civil transport aeroplane is a particular class of aircraft.

The features which distinguish it from other aircraft are the following:

1. It is a *civil* aircraft, in contrast to military aircraft;
2. It is an *aeroplane*, in contrast to other contrivances capable of moving in the atmosphere;
3. It is or can be used for *transportation*, in contrast to other possible uses.

CIVIL AND MILITARY AIRCRAFT

A first attempt to classify aircraft along these lines was made by Fauchille, [1] who made a distinction between public and private balloons. He regarded military balloons as being public balloons.

A similar classification is to be found in the Paris Convention, Art. 30 of which states that:

"Seront considérés comme aéronefs d'Etat:

a. Les aéronefs militaires;
b. Les aéronefs exclusivement affectés à un service d'Etat, tel que: Postes, Douanes, Police.

Les autres aéronefs seront réputés aéronefs privés. Tous les aéronefs d'Etat, autres que les aéronefs militaires, de douane ou de police, seront traités comme des aéronefs privés et soumis, de ce chef, à toutes les dispositions de la présente Convention".

Art. 3 of the Chicago Convention also contains the following provision:

"*a.* This Convention shall be applicable only to civil aircraft, and shall not be applicable to state aircraft.
b. Aircraft used in military, customs and police services shall be deemed to be state aircraft".

In view of the foregoing, it is apparent that we can approach our objective of defining "civil aircraft" by endeavouring to

---

[1] Annuaire de l'Institut de Droit International 1902, page 25.

establish what is meant by military aircraft. The other categories of State aircraft, i.e. customs and police aircraft, are comparatively easy to identify as such, while their numbers have hitherto been extremely limited [1].

Postal aircraft, which the Paris Convention classed as State aircraft but treated as private aircraft — a construction which is not beyond challenge and has therefore been the subject of considerable criticism [2] —, are no longer mentioned as a separate category in the Chicago Convention.

What exactly is a military aircraft? From the number of definitions that have been formulated in the course of time, it is obviously no easy matter to answer this question. Zondag [3] lists the following criteria which may be applied: the ownership, the armament or intention to be used in war, military command, military command and the fact of being attached to the armed forces, military markings, use for military purposes, the fact of forming part of the military strength, or a combination of these criteria. Cooper [4] also mentions the criterion of being under the command of an officer in uniform and carrying a military certificate.

The existing Dutch Aviation Act does not define the term "military aircraft". In the explanatory memorandum appended to this Act it is stated that such a definition would be out of place in the Aviation Act. According to Art. 50 (m) the entry of military as well as police and customs aircraft of foreign nationality must be regulated by general Order. This was done in the Decree of April 29, 1931, No. S. 179, Art. 1a of which gives the following definition of foreign military aircraft: "All aircraft of foreign nationality commanded by a person in foreign military service detailed for the purpose by the competent authority". Many objections have been raised against this definition, which is copied word for word from Art. 31 of the Paris Convention. It is pointed out that it is incorrect to determine the military character of an aircraft solely by the nature of the command, i.e. the personnel element, while the nature of the aircraft itself is igno-

---

[1] For the increasing use of aircraft by police authorities in America see Ogburn, "The Social Effects of Aviation" page 434 onwards.

[2] See Goedhuis, "Handboek voor het Luchtrecht" page 61.

[3] Zondag, "Neutraliteit in de lucht" page 10 onwards.

[4] Cooper, "The Legal Status of aircraft" page 31.

red. [1] However, it is also questionable whether the nature of the aircraft would be a serviceable criterion. As Parker van Zandt [2] observes: "No civil plane is too slow or small to serve some military purpose." The definition in Art. 3 of the Chicago Convention is therefore more satisfactory, since the *use* of the aircraft expressly determines whether it must be regarded as a military, customs or police aircraft [3].

Cooper, [4] who was largely responsible for the wording of this Article, remarks that the definition was deliberately kept more or less vague with the object of obtaining a more practical solution than had previously been possible with more precise definitions. Actually, however, the difficulty is merely shifted, because the court in each country concerned will now have to decide what is meant by "military use" when the case arises.

There can be little doubt that an aircraft carrying a company of parachute troops, for example, is being used for military purposes. But where is one to draw the line with a commercial plane whose passengers include a number of high-ranking officers on their way to attend a military conference? As air transport has come to be regarded as an integral part of every air force, [5] this gives rise to a number of problems which are beyond the scope of the present study.

THE AEROPLANE FROM A TECHNICAL ASPECT

In Annex 7 of the Chicago Convention, aircraft are classified according to their technical characteristics, a distinction being made between aircraft lighter than air and those heavier than air. Both categories are subdivided into power-driven and non-power-driven aircraft. The group of heavier-than-air power-driven

---

[1] Goedhuis, op. cit. page 62, Zondag op. cit. page 13.

[2] Parker van Zandt, "Civil Aviation and Peace" page 23.

[3] Similarly in the project for a new Netherlands Aviation Act (article 17), all aircraft used for military purposes are understood to be military aircraft.

[4] "The Legal Status of Aircraft", pages 51 and 52.

[5] For the use of transport aircraft for military purposes see Parker van Zandt, "Civil Aviation and Peace" page 22 onwards; Reginald M. Cleveland "Air Transport at War"; Major Oliver Stewart, "Air Power and the expanding community" page 187; Craven, "The Army Air Forces in World War II", Part I, page 310 onwards; "Survival in the Air Age" a report by the Presidents Air Policy Commission, page 36; Wolfe, "Air Transportation-Traffic and Management" page 134; Wright, "Aviation's Place in Civilization" page 15.

aircraft is found to include aeroplanes, gyroplanes, helicopters and ornithopters.

Both the gyroplane and the ornithopter can be ignored, as they are really outmoded and are no longer produced.

The helicopter, however, will probably be used to an ever-increasing extent. In the existing Netherlands Aviation Act, and also in the slightly revised text of the proposed new Act, the entire group of "mechanically-propelled heavier-than-air craft" are referred to as "aeroplanes". It must be borne in mind that this includes both conventional aircraft (aeroplanes) and helicopters. In our opinion it would be better if a distinction could be made between these two categories of aircraft in the Aviation Act, on account of their entirely different characteristics — one need only think of the helicopter's ability to hover in the air and to land practically anywhere. Such a distinction has been made in the R.T.L. with regard to autogyros [1].

There is still another distinction, based on the element from which the aircraft takes off and on which it lands, viz. a subdivision into landplanes, seaplanes and amphibians. This raises the question of whether an aircraft on the water ought to be regarded as a ship.

The general opinion seems to be that this is not the case [2], though certain rules in force for shipping also apply to aircraft on the water. In Art. 66 of the Netherlands Air Traffic Regulations, for example, the Inland Waterways Collision Rules have been declared applicable.

A similar question arises with regard to landplanes on a public highway, i.e. not at an airport. This is a point of practical importance since there are "roadable aircraft" which can easily be transformed into vehicles resembling automobiles. Such vehicles must not be simply classed as automobiles, though it will be necessary to declare the traffic regulations applicable to this type of vehicle [3].

---

[1] RTL article 1, sub II and sub III.

[2] Cf. Goedhuis, ,,Handboek voor het Luchtrecht" page 60; Oppikofer, "Das Sonderrecht des Wasserflugzeuges", Archf LR 1934, page 74; Dijkstra, "Business Law of Aviation" page 3 onwards; see also Rhyne, "Aviation Accident Law" page 282.

[3] The U.S.A. National Motor Vehicle Theft Act has been declared applicable for "any contrivance for flight in the air", see Dijkstra, op. cit. page 4.

USE FOR TRANSPORT PURPOSES

The manner of use was another criterion mentioned for the transport aeroplane. Although commercial air transportation is undoubtedly the commonest form of air navigation, one must not forget that this is only one of many uses. Aircraft are or have been employed for spraying crops, aerial mapping, destruction of insects, training purposes, detecting shoals of fish, mineralogical surveys, meteorological observations, ambulance and rescue work, inspection of cattle herds and pipe lines, combating forest fires, causing rain to fall, recreation and advertising purposes, to mention only a few of the possible uses [1].

The number of civil transport aircraft in operation all over the world is comparatively small. The airlines which are members of IATA carry 95% of all the regular air traffic and they have a total fleet of about 2.500 aircraft [2].

As a definition of a transport aircraft we shall adopt the description given in the R.T.L., viz.: "Aircraft serving for the transportation of persons and/or goods as a business or as part of the business" [3].

The use of an aircraft as a transport aircraft is subject to a formal requirement, in that the captain is forbidden to use an aircraft for transportation purposes if the certificate of airworthiness is not endorsed to the effect [4].

Both the Dutch legislator and ICAO have laid down stricter regulations for the use of transport aircraft than for other forms of air navigation [5].

In commercial air transportation a distinction must be made between scheduled and non-scheduled operations. From the viewpoint of international air policy the distinction is very important, and ICAO only recently succeeded in formulating a satisfactory definition of a "scheduled international air service." [6]

For the aircraft commander, however, the difference is not of such fundamental importance, since ICAO has decided in princi-

---

[1] See Ogburn, "The Social Effects of Aviation"; Wright, "Aviation's Place in Civilization" page 21.
[2] "Avia Vliegwereld" 28th August, 1952, page 483.
[3] RTL article 1, sub. VI.
[4] RTL article 189.
[5] Cf. RTL, Chapter XI and ICAO Annex 6.
[6] ICAO Doc. 7278–C/841.

ple that an equal level of safety must be maintained in both scheduled and non-scheduled operations [1].

However, owing to the different character of the two forms of operation, it is often impossible to apply identical rules in both cases. In scheduled operations, for example, weather minima [2] have to be fixed for all the airfields used, whereas in non-scheduled operations it is sufficient to specify a method of establishing the weather minima [3].

This difference is of course explained by the special character of non-scheduled operations, which makes it impracticable to state beforehand the airfields that will be used and the routes to be flown.

Scheduled carriers have also started operating non-scheduled flights. In the case of K.L.M. these "special flights" constitute an important part of the company's business. The K.L.M. captains usually perform scheduled flights, but from time to time they may be assigned to non-scheduled commercial flights.

NATIONALITY AND REGISTRATION

Aircraft, like other means of transport such as ships, trains and automobiles, are able to cross the national frontiers. The latter vehicles usually bear a distinguishing mark of the country of origin. What is the connection between the vehicle and the country concerned? In the case of ships and aircraft, the connection is often referred to as "nationality".

In applying this term to aircraft, however, certain writers arrive at widely differing conclusions.

Riese [4], for example, describes the use of the term "nationality" with regard to aircraft as "durchaus unklar und irreführend" [5]. He rejects comparison with the "law of the flag" of seagoing vessels, because the latter normally move in stateless territory, whereas aircraft on overland flights fly above the territory of a country or countries, which leads him to the conclusion that:

---

[1] Final Report of the Operations Division, 3rd Session, Doc. 6640–OPS/567, page 28.

[2] For the significance of weather minima for safety see page 61 onwards.

[3] Annex 6, par. 4.2.5. Similar differences are evident when establishing minimum altitudes (Annex 6, par. 4.2.4.) and the experience requirements for pilots-in-command (Annex 6, par. 9.1).

[4] Riese, „Luftrecht" page 202 onwards.

[5] Also Mandl, "La "Nationalité" des Aéronefs n'est qu'une dénomination erronée", Droit Aerien 1931, page 161.

"Insofern ähnelt die Lage des Flugzeugs im Überlandverkehr mehr der des Automobils oder des Binnenschiffes als der des Seeschiffes". On the other hand Goedhuis observes that "aircraft require a nationality and there is no essential difference between the nature of this nationality and that of a ships's nationality." [1] Cooper [2] likewise submits that "... aircraft, like vessels, and unlike railway trains and automobile vehicles, now have that quality of legal quasi-personality in public international law discussed above as *nationality*", though he attributes a limited significance to the term.

We prefer the latter view, which is based on a historical analysis of the term „nationality" in maritime law and air law. It is concise, and based on the following considerations. An aircraft used for international transportation must be under the control of a State, which will see to it that the aircraft in question fulfils its obligations while in other States; this task devolves upon the State of origin of the aircraft. At the same time, however, this State naturally acts as protector of the rights of the aircraft concerned. This limited conception of the term "nationality", which definitely does not have any significance in civil or criminal law, was formulated by Kriege as early as 1910 [3]. It appears to be confirmed by actual practice in international aviation as well as by sundry provisions of international air agreements.

The State's obligation to control its national aircraft wherever they are, is expressly mentioned in Art. 12 of the Chicago Convention, viz.:

> "Each contracting State undertakes to adopt measures to insure that every aircraft flying over or manoeuvring within its territory and that every aircraft carrying its nationality mark, wherever such aircraft may be, shall comply with the rules and regulations relating to the flight and manoeuvre of aircraft there in force".

And in Art. 5 of the same Convention, for example, certain rights are granted to "all aircraft of the other contracting States". The wording employed appears to confirm the view that the States concerned can stand up for their national aircraft should

---

[1] „Handboek voor het Luchtrecht" page 65.
[2] Cooper, "The Legal Status of Aircraft" page 61.
[3] See footnote in Cooper, op.cit. page 27.

the rights granted be infringed. An example of this occurred in 1948 when the Netherlands protested to India against the ban on K.L.M. aircraft flying over Indian territory.

Nationality is therefore of importance for the aircraft commander because it means that all over the world he is subject to the control of the State where his aircraft is registered, and also because he can count on protection by this State if the rights granted to his aircraft are infringed. No other significance should be attached to the term "nationality."

Another point worth noting is that as a general rule it is not obligatory for the aircraft commander — unlike the master of a ship — to be of the same nationality as the aircraft.

This requirement, which is mentioned even by Fauchille, has never found acceptance in Dutch aviation [1]. In the air agreement concluded between the Netherlands and Brazil, nevertheless, Art. 4 provides that the rights granted under the terms of the agreement may be withdrawn or withheld if the aircraft of the other party are not manned by a crew whose members are nationals of that contracting party, except when aircrew personnel are being trained.

Of the 517 pilots employed by K.L.M. in 1952, 174 were of non-Dutch nationality [2]. In this case the employment of foreign personnel is subject to the consent of the Minister of Transport & Waterways.

Under the Chicago Convention the nationality of an aircraft is determined by a formal criterion, namely the registration. According to Art. 17 of this Convention, "Aircraft have the nationality of the State in which they are registered." Aircraft cannot be validly registered in more than one State, but the registration can be transferred from one State to another (Art. 18).

The manner of registration or transfer is left to the national legislators. Most of the existing rules on the subject are still based on the original text (cancelled in 1929) of the Paris Convention, which laid down certain requirements with regard to the

---

[1] It exists in French Legislation, see Maschino, "La Condition Juridique du Personnel Aérien" pages 34 and 35. For the difficulties which may arise from such a ruling see Goedhuis, „La Situation Juridique du Commandant de l'Aéronef", RDILC 1933, page 137.
[2] Information provided by K.L.M.

nationality of the owner himself or of the persons who form the
Board of Directors of the airline concerned [1].

JURISDICTION ON BOARD

What law applies to occurences and transactions on board an
aircraft in flight? This is a question which has absorbed the
attention of many different writers [2]. As Lemoine remarks, it is
"un problème complexe et difficile. Il n'est que très fragmentaire-
ment résolu par le droit positif et les différents systèmes nationaux
n'offrent pas tous des solutions concordantes. C'est dire qu'il
permet des discussions nombreuses et, de fait, les auteurs qui
s'y sont appliqués en proposent des solutions variées. " [3]
The scope of the present study naturally imposes limitations on
the extent to which this old "crux jurisconsultorum" [4] can be
explored. We shall therefore ignore important aspects such as the
differentiation between civil law and criminal law, the question of
whether a transaction on board has consequences outside the
aircraft or not, the situation in the case of flights over „stateless"
territory, etc. — and merely state that in general there are five
possible systems, all of which have their supporters [5].
   1. The theory of territoriality. According to this system the
only law applicable is the law of the State over whose territory the
aircraft is flying. Objections are that the position of an aircraft is
sometimes difficult to determine, and that an aircraft flying over
the sea or above stateless territory would be in a legal vacuum.

---

[1] Cf. Goedhuis, „Handboek voor het Luchtrecht" p. 66; Dutch applicable legislation is to be found in RTL art. 3–11.
   [2] Niemeyer, "Crimes et délits commis à bord des aéronefs", R.D.A. 1929, page 285.
   Pholien, "Des crimes et délits commis à bord d'aéronefs en vol", R.D.A. 1929,
page 289.
   Meyer, "Crimes et délits à bord des aéronefs", R.G.D.A. 1946, page 544.
   de Visscher, "Le Règlement des compétences pénales en Droit Aérien", R.G.D.A.
1937, page 329.
   Idem, "Les conflits de lois en matière de droit aérien". Recueil des Cours de
l'Academie de droit international à la Haye 1934, T. II, page 279.
   Volkman, "Crimes et délits à bord des aéronefs en droit international", Droit
Aérien 1931, page 26.
   Danilovicz-Szondy, "Les infractions à la loi pénale commises à bord des aéronefs",
Droit Aérien 1930, page 402.
   Cooper, "The legal status of aircraft".
   Lemoine, "Traité de droit aérien" pages 201 and 792.
   [3] Lemoine, op cit. page 201.
   [4] Volkman, loc. cit. page 26.
   [5] Lemoine, op. cit. page 202; Meyer, loc. cit. page 614; Danilovicz, loc. cit. also
mentions 5 systems but based on different criteria.

2. The theory of nationality. In this case the only law applicable is the law of the country whose nationality the aircraft possesses. One of the criticisms raised against this system is that the theory does not take full account of the sovereign rights of the States traversed by the aircraft.

3. and 4. The theories of the competency of the place of departure and the place of arrival. An objection against both theories is that they contain an arbitrary element, and· that countries declared competent may not have the slightest interest in the occurrence in question.

5. Lastly, there are a number of combinations of the above theories, usually based on the very general assumption that the nationality principle applies in the first place, but that in criminal proceedings a State flown over by the aircraft is likewise competent insofar as its interests are directly at stake; naturally the maxim of "non bis in idem" applies in this connection.

In positive law a similar "mixed" construction of this nature is to be found in Swiss, French and German legislation [1]. The nationality principle appears to predominate in the Italian legislation [2].

The territoriality principle still prevails in Dutch legislation. Criminal offences committed on board Dutch aircraft outside the Netherlands are not indictable in the Netherlands; above Dutch territory Dutch law applies to all aircraft, including those of foreign nationality [3].

In addition to the writers already mentioned, several bodies such as the International Law Association, the Comité Juridique International de l'Aviation, Harvard University and the Institut de Droit International, [4] have formulated proposals which envisage a uniform solution for this problem. At the moment there is still no question of unity of ideas, much less of legislation. Although this is undoubtedly a deficiency which will have to be made good, the lack of a generally applicable regulation on the subject has hitherto caused little difficulty in actual practice. The reason for this is probably the fact that transactions (drawing

---

[1] Meyer, loc. cit. page 619.
[2] Ibidem page 621.
[3] See page 76.
[4] See Cooper, op.cit. Appendix A; I.L.A. Air Law Committee, Lucerne Conference 1952, page 2.

up contracts) and occurrences (births and deaths) to which civil law would be applicable, have so far been comparatively rare on board aircraft. The same applies with regard to criminal offences committed on board aircraft in general [1].

Moreover, there is considerable agreement on the question of competency in regard to those criminal offences which are perhaps most likely to occur on board aircraft, viz. infractions of air navigation regulations and breaches of discipline.

Even the greatest advocates of freedom in the air have always recognized the right and also the obligation of a State to control the use of the atmosphere above its territory. [2] On the other hand, it is generally accepted that a State is likewise obliged to control the actions of its national aircraft, wherever they may be. International agreement has been reached on both principles, as may be seen from Art. 12 of the Chicago Convention. [3]

With regard to breaches of discipline on board the aircraft, it is fairly generally agreed that these must be judged solely by the State whose nationality the aircraft possesses. [4]

[1] For such an occurrence see page 79.
[2] Lemoine, op. cit. page 792.
[3] See page 30 and page 48.
[4] To this effect: de Visscher, R.G.D.A. 1937 page 342 and Receuil des Cours 1934 page 378; Meyer, loc. cit. page 625; proposal of the ,,Instituut de Droit International'' quoted by Cooper, op. cit. page 68.

CHAPTER III

THE CREW

THE AIRCRAFT COMMANDER AS A MEMBER OF THE CREW

Although the Netherlands Aviation Act contains numerous references to the „commander" of an aircraft [1], it does not specify who is meant by this term.

From Article I (V), however, it follows that in the first place he is a member of the crew, since that article gives the following definition for a "member of the crew of an aircraft":

"the commander and any other person who has to perform duties in an aircraft during and in connection with its flight".

According to the explanatory memorandum [2] appended to the Aviation Act, this definition was worded in very general terms so that it would cover not only the pilots but also the flight engineers and radio operators, amongst others. Although this enumeration of categories is not limitative, it may be deduced from the explanation as well as from the above definition itself that really only the technical personnel were meant to be classed as crew members. Art. 15 of the Aviation Act and Art. 19 of the R.T.L. jointly designate the crew members who are regarded as belonging to this category.

Art. 15 (1) of the Aviation Act states that it is forbidden for a member of the crew to do duty on board an aircraft unless he holds a valid certificate of competency for that duty, while Art. 19 of the R.T.L. states that as far as the performance of air navigation with aeroplanes and autogyros is concerned, the certificates of competency referred to in Art. 15 are classified as follows:

1. Pilot's licence;
2. Navigator's licence;

---

[1] See page 46.
[2] M. v. A. 1921.

3. Flight Engineer's licence;
4. Radio Telephonist's licence;
5. Radio Telegraphist's licence [1].

This shows which personnel are to be looked upon as members of the crew; it will be noticed that there is no special licence for the aircraft commander. In the case of other categories of personnel who may perform duties on board the aircraft, e.g. stewards, stewardesses, cooks, flight clerks, aerial photographers, technicians carried on test flights, inspectors, examiners, nurses, special stewards to look after cargoes of animals, etc., it cannot be claimed that they perform their duties "in connection with the flight".

Accordingly, they are not members of the crew within the meaning of the regulation in question and so they do not need to be in possession of a certificate of competency as a member of the crew, as prescribed in Art. 15 of the Aviation Act.

This limited conception of the term "member of the crew" agrees with the definition of "flight crew members" in various ICAO regulations [2], viz.:

"A licensed crew member charged with duties essential to the operation of an aircraft during flight time".

In the same ICAO regulations, however, the term "crew" is given a wider significance than the term "flight crew", viz.: "Those persons assigned by an operator to duties on an aircraft during flight time" [3].

The latter definition thus applies to categories of flying personnel who are at present completely ignored by the Netherlands Aviation Act. A similarly wide interpretation of the word "crew" also occurs in Dutch legislation, however. Art. 1 (d) of the Act of October 26, 1935 (S. 626) — the Act for the regulation of the Sanitary Control of Aerial Navigation — defines "crew" as meaning "each person who is designated by or on behalf of the owner or holder of an aircraft to perform services on board that aircraft".

In conclusion it may be mentioned that the definition of the

---

[1] With regard to balloons mention is made in article 62 RTL of licences to fly as a balloonist.

[2] For example Annex 9, "Facilitation of Air Transport", Chapter 1, Definitions.

[3] For the notions "crew" and "flightcrew" in French and Italian air legislation see Bucher, op. cit. page 18. For the Swiss regulations see Riese, "Luftrecht" page 211.

term "member of the crew" will probably be omitted from the new Dutch Aviation Act (by avoiding the use of this expression in the text altogether).

Does the Dutch system of legislation indicate that the captaincy attaches to a particular function on board the aircraft?

Art. 70 of the R.T.L. stipulates that if the crew of an aircraft consists of more than one member, the operator must designate one of the members as the commander. This means that the function of aircraft commander can only be exercised by the categories classed as "members of the crew", but that in principle the operator is otherwise free to make his own choice. He can therefore appoint a pilot, navigator, flight engineer, radio telephonist or radio telegraphist as commander, though under the terms of Art. 15 of the Aviation Act as a general rule they must be in possession of a valid certificate of competency. [1] The holder of the aircraft is not allowed to choose a steward as commander, for example, since stewards are not included in the above categories.

In actual practice one of the pilots is always designated as commander on a Dutch aircraft. This is also in accordance with recent ICAO regulations, in which the term "aircraft commander" has now been superseded by "pilot-in-command." The intention is that the new Dutch air legislation will likewise establish the principle that only a pilot can be designated as aircraft commander.

Another possibility is quite conceivable, however.

In the French act of May 31, 1924, for example, it is stipulated in Art. 31 that the aircraft commander must have a special licence as "commandant". The regulations issued under this Act on February 10, 1926, do not go into further detail, but Maschino [2] is of the opinion that a holder of a 1st Class Navigator's licence is the aircraft commander.

In the Netherlands Naval Air Service there were also cases where a navigator other than the pilot was in command of a flying boat. Art. 45 of the R.T.L. shows that the Dutch legislators likewise envisaged the possibility of a navigator acting as aircraft commander.

---

[1] See page 38 for an exception hereto.
[2] Maschino, "La Condition Juridique du Personnel Aérien" page 93.

This article which describes the duties of the navigator, states that *"if he is not at the same time the commander"* [1] the navigator has to give the commander advice on all matters relating to the navigation.

Art. 4 of the Aviation Act makes provision for cases where the crew only consists of one person. Naturally the person in question is a pilot, and the Article prescribes that insofar as the regulations contained in the Aviation Act are applicable to the commander of an aircraft they shall also be applicable to any person operating an aircraft alone.

In our further observations we shall confine ourselves to the case of a pilot acting as aircraft commander. Other important Articles relating to the position of the aircraft commander as a member of the crew of a transport aircraft are Arts. 67 and 68 of the R.T.L. These deal with the composition of the crew, depending on the nature and length of the route and whether the flight is performed during the day or during the night.

It is forbidden to serve as commander on board a transport aircraft if the above regulations are not satisfied in respect of that aircraft (Art. 70. 4, R.T.L.). As the regulations are purely technical in addition to being very much out-of-date, we shall not devote any more time to their contents.

CERTIFICATES OF COMPETENCY

We have already seen that the aircraft commander, as a member of the crew in general, must hold a valid certificate of competency for the duties which he is to perform on board the aircraft.

According to current Dutch law [2], however, an exception is made for the performance of certain activities under the immediate supervision of the holder of a valid certificate of competency for those activities, if the latter is in a position to intervene at all times.

This provision is naturally intended to make it possible for student pilots, for example, to acquire experience in handling aircraft under the supervision of an instructor. In theory, the operator can appoint the student pilot as commander in such a

---

[1] The italics are ours.
[2] Aviation Act, Article 15 par. 3a.

case — though it may be assumed that naturally he will think twice before doing so. But the present wording of this exceptive clause opens up the following possibility which is quite conceivable in actual practice: a very experienced senior pilot may lose his licence for certain reasons, but nonetheless the operator may appoint him to take general control of the flight in the capacity of aircraft commander. Insofar as he takes an active part in the piloting of the aircraft, however, he can only do so under the supervision of a licensed pilot who must be able to intervene at all times.

Both ICAO and IATA have adopted the standpoint — and rightly so, in our opinion — that it is inadvisable to introduce the figure of a "non-piloting aircraft commander", partly in view of the fact that "the presence on board of an experienced pilot no longer licensed to serve in a piloting capacity might lead to 'cockpit trouble', since it would be difficult for him to resist taking over the controls in an emergency." [1]

Art. 15 (Par. 3b) of the Aviation Act makes provision for other exceptions, viz. insofar as exemption has been granted by or on behalf of the Minister of Transport, as well as in cases specified by general order. In this connection Arts. 196 ter and quater of the R.T.L. state that the prohibition contained in Art. 15 (1) of the Aviation Act shall not apply to training flights and examination flights, provided that these are held within a distance of 3 kilometres from the boundary of an aerodrome which may be used for such flights, and provided that the pilot — subject to exemption by or on behalf of the Minister — satisfies the physical fitness requirements for the licence in question. The distance of 3 kilometres mentioned in this Article is generally inadequate for larger and faster aircraft. Moreover, for safety reasons it may be most undesirable to perform such flights in the immediate vicinity of an aerodrome. This regulation is therefore out-of-date and one may assume that a regulation more appropriate to modern air navigation will replace it in the new R.T.L.

Under the terms of Art. 16 of the Aviation Act, certificates of competency are issued, suspended and withdrawn by or on behalf of the Minister of Transport. These powers are entrusted to and

[1] Final Report of the Personnel Licensing Division 3rd Session, Doc. 5408–PEL 535, page 20; an opposite view is presented by Knauth, JAL 1947, p. 161.

exercised by the Director-General of the Netherlands Department of Civil Aviation. What are the guiding principles adopted in this connection? Chapter IV of the R.T.L. gives detailed but now obsolete regulations on the subject. Candidates must undergo a medical examination and periodical re-examinations in addition to passing a theoretical and practical examination. Certain age limits are also laid down for the first issue of the pilot's licence, viz. at least 18 years for the A Licence ("private pilot") and at least 21 years and not more than 35 years — subject to exemption — for the B Licence ("airline pilot").

According to the explanatory memorandum [1] appended to the Aviation Act, the original intention was that eligibility for a licence should be judged in the light of qualities of character, etc., as well as the technical skill and physical fitness of the candidate. This aim was not achieved in the drafting of the R.T.L., since Art. 20 states that a licence will be issued "to each applicant, provided that he satisfies the requirements laid down in this Chapter with regard to the licence applied for and provided that the remaining provisions of these regulations are satisfied". Requirements or regulations concerning "qualities of character, etc.", however, are not to be found in the R.T.L.

Under Dutch law it is therefore impossible to refuse to issue a licence to a candidate on other grounds if he satisfies the requirements laid down in the R.T.L. [2].

This is in contrast to the practice in the U.S.A. where, for example, no licence was granted to a candidate who had shown that he possessed ample technical skill as a pilot and who satisfied all the other requirements, but who was not considered to be "properly qualified" in view of the fact that he had a large number of convictions for breaking road traffic regulations [3]. Qualities of character can also be taken into consideration in the United Kingdom [4] and in Switzerland [5] when deciding whether a certificate of competency shall be issued.

Needless to say, it is of great importance for air safety that

[1] M.v.A. 1912.

[2] However, in such a case action can be taken by the Air Accident Board, see page 82.

[3] Merill Armour, "Analysis of the CAB Precedents in Safety Enforcement Cases", JAL 1950, page 54.

[4] Air Navigation Order 1949, par. 20. 10.

[5] Cf. Riese, "Luftrecht" page 214.

emphasis should be laid on those qualities which distinguish a safe pilot from an "unsafe pilot" during the medical examinations, written examinations and flight tests, and when selecting flying personnel in general. From a report [1] published by the C.A.A. concerning a survey carried out by the National Research Council Committee on Aviation Psychology, there appears to be some doubt as to whether this is always the case, in the United States at least:

"Data obtained in this survey indicate that many of the most critical components of the pilot's job are not considered critical by the captains and Civil Aeronautics Administration examiners who evaluate other pilots. There is some evidence that these check-pilots emphasize components of the job which seldom contribute to critical situations and accidents".

The circumstances under which a pilot has to perform his duties as well as the nature of these duties, make his mental disposition very important [2]. Some people are therefore in favour of making a psychiatric examination and periodical re-examinations obligatory for airline pilots [3]. At the fourth meeting of the Operations Division of ICAO in Montreal, Dr McFarland likewise drew attention to the desirability of pilot selection on a psychological basis. [4]

On the basis of a critical examination of the numerous publications on this subject, British investigators [5] have come to the conclusion that "although there is now general agreement that some form of psychological testing is desirable its form is difficult to define."

The conclusion reached by another investigator [6] who carried out extensive selection tests with R.A.F. pilots and who was in a

---

[1] Thomas Gordon, "The Airline Pilot: a survey of the critical requirements of his job and of pilot evaluation and selection procedures", page 54.

[2] Cf. Mamoru Mochizuki, "Aviation Psychology"; Tweedie, "Air Accidents Investigation and Human Failure"; van Lennep, "Experiences in connection with the Psychology of the Commercial Pilot"; van Balkom "Experiences with regard to the psychology of pilots"; Lectures for the Royal Swedish Academy of Engineering Sciences, Stockholm 1951; Bond, "The Love and Fear of Flying".

[3] Dr. Maurice M. Walsh, at the 21st Annual Conference of the Aero Medical Association at Chicago, Interavia News Letter No. 1975, 10th June, 1950, page 4.

[4] Appendix A to Doc. 7151 — OPS/609 pages 135 and 136.

[5] Air Vice Marshal Sir Charles P. Symonds and Wing Commander Denis J. Williams, "Psychological Disorders in Flying Personnel of the Royal Air Force" page 11.

[6] D. Russell Davis, "Pilot Error, Some Laboratory Experiments" page 37.

position to compare the results of these tests with the subsequent flying career of the pilots tested, is as follows:

"It was found that those who in the test showed the inert reaction in a marked form were more often suspended from training than other subjects, sustained more fatal accidents and more often became operational casualties. Those who showed errors of preoccupation in a marked form were more often suspended from training than others."

In the Netherlands, psychological tests have been employed in the selection of airline pilots for some time past, but an unfavourable report need not prevent a candidate from securing an airline pilot's licence if he otherwise meets the official requirements. It is quite possible that further research in the methods of selecting flying personnel will yield important results for air safety.

AUTHORITY OVER THE PERSONS ON BOARD

According to Dutch maritime law, the master of the ship exercises authority over all on board [1].

Various foreign air legislations contain similar provisions with regard to the commander of an aircraft, e.g. the Brazilian Act of June 8, 1938 (Art. 151), the French Act of March 25, 1936 (Art. 8), the Italian Act of February 8, 1934 (Art. 17), the Czechoslovakian Act of July 8, 1925 (Art. 16), the Russian Act of August 7, 1935 (Art. 23) and the Argentine draft of a Civil Aviation Code, 1938 (Art. 75) [2].

In Dutch air legislation, however, there are no general provisions of like tenor. Although the Aviation Act and the orders issued under that Act make numerous references to the „commander" [3], from which it may be inferred that the legislator is speaking of the person in command of the aircraft, the principle is not definitely established.

The power to exercise authority is indirectly expressed in Art. 2 (1) of the Netherlands Air Traffic Regulations, which reads as follows: "The commander of an aircraft is responsible for compliance with the provisions of these regulations and is obliged to take all steps necessary for this purpose".

---

[1] Article 341 W.v.K.
[2] ICAO Doc. 4538/LC 18.
[3] See page 46.

Although this Article does not specifically mention authority over the persons on board, in our opinion it does imply a very far-reaching authority; if the aircraft commander considers such action necessary to secure compliance with the provisions of the Air Traffic Regulations, he may take steps to compel those on board to obey his orders. The reasoning behind this regulation is obvious. If the aircraft commander is held responsible for compliance with certain instructions or for the safety of the flight in general, he must be furnished with the powers to acquit himself of this responsibility.

As the legislator has accepted this principle in the Air Traffic Regulations, one wonders why a similar provision is not included in the R.T.L. and in the Aviation Act itself, which likewise impose numerous obligations on the commander. Here we have the remarkable situation of the aircraft commander being legally authorized — in virtue of Art. 2 of the Air Traffic Regulations — to employ force if necessary in dealing with a radio operator who refuses to transmit a certain signal [1] or a passenger who shines a dazzling light [2] or wishes to drop things out of the aircraft [3].

On the other hand, no such power is granted to an aircraft commander who wishes to take action against a passenger who smokes in a place where this is forbidden [4] or who tries to force his way into the cockpit [5].

We consider that in Dutch air legislation there ought to be a general provision to the effect that the commander of an aircraft exercises authority over the persons on board; this is a deficiency which must be made good. In the draft convention on the legal status of the aircraft commander an attempt is made to lay down rules on the subject [6].

---

[1] Article 14 LVR.
[2] Article 3 LVR.
[3] Article 64 LVR.
[4] Article 194 RTL.
[5] Article 193 RTL.
[6] See page 135.

CHAPTER IV

FLIGHT OPERATIONS

An aircraft on the ground is "a useless, expensive, defenseless contraption", but in the air it is "a thing of beauty and surprising power."[1] The purpose of an aircraft and crew is to fly; only during flight can the combination of aircraft and crew be productive.

In the preceding chapters we have briefly discussed the aircraft and crew elements. We shall now examine the rules which the aircraft commander must observe in preparing for a flight as well as in its execution.

PREPARATION FOR A FLIGHT

ICAO has drawn up a number of standards which the commander of a transport aircraft must satisfy before commencing a flight. Prior to the flight, the pilot-in-command has to sign a form in which he declares that he is satisfied that the aircraft is airworthy, that the necessary equipment is on board, that a maintenance release has been issued, that the weight of the aircraft is such that a safe flight is possible under the anticipated conditions, that the cargo is distributed and secured in such a way that the aircraft is safe for the flight, that the regulations concerning performance requirements have been complied with and that the flight plan has been drawn up in the prescribed manner and duly signed. These forms must be kept by the operator for a period of six months[2].

Two important conclusions with regard to the aircraft commander can be drawn from this regulation.

In the first place the aircraft commander is required to state

---

[1] Air Marshall Sir Arthur Coningham quoted in "Orgaan van de Vereniging ter beoefening van de Krijgswetenschap" 1951–1952, page 140.

[2] Annex 6 par. 4. 3.

that he is satisfied that the points specifically mentioned are in order. What are we to understand by the words "is satisfied"? It would be going too far, besides being practically impossible, to insist that the commander should satisfy himself personally that the aforesaid requirements have been met. The technical side of aviation has become too specialized and transport aircraft are too complex for the commander to be able to carry out a personal investigation of the airworthiness or to check the stowage of the cargo. The wording of the above regulation implies that it is left to the commander to decide how he wishes to satisfy himself that everything is in order; he can depend on statements by third parties, for example, or make random checks himself. The system thus permits delegation of inspection duties by the aircraft commander to other members of the crew or to ground personnel, though the aircraft commander has the final say and also bears the responsibility.

A second important point is that the above regulation *indirectly* entitles the aircraft commander not to commence a flight if he is of the opinion that it cannot be safely performed. The flight plan, which must be signed for approval by the commander — and also by the flight operations officer, where applicable [1] — before the flight, should indicate that the flight can be conducted with safety. [2] By refusing to sign the flight plan on grounds of air safety, the aircraft commander can legally shield himself from any obligation which the operator may impose upon him to commence the flight. A refusal to depart under such circumstances does not constitute a breach of a contract of employment between the two parties [3].

In our opinion this right of refusal to depart is inherent in the commander's responsibility for the aircraft and the persons on board during the flight. One can hardly expect him to commence a flight and thus put himself in a situation which he deems beforehand to be unsafe and therefore not in keeping with his responsibility.

The French legislation expressly stipulates that the departure

---

[1] See page 57.
[2] Annex 6 par. 4. 3. 1.
[3] The same applies to the ship's captain, see Cleveringa, "Het Nieuwe Zeerecht" page 236.

may only take place "avec l'accord du Commandant" [1]. There seems to be a historic reason for this provision, however, as the opposite standpoint had originally been adopted [2]. In view of the foregoing, it appears to us that such a declaration is now super-fluous.

We shall not go into the other detailed regulations which the aircraft commander has to comply with before departure, e.g. the regulations concerning the weather conditions, [3] the quanti-ties of fuel and oil to be carried [4], and the oxygen equipment [5].

EXECUTION OF A FLIGHT

The guiding principle in regard to the responsibility of the aircraft commander during the flight is that:

"The pilot-in-command shall be responsible for the operation and safety of the aircraft and for the safety of all persons on board, during flight time." [6] The wording of this principle will be studied in further detail in Part II [7], when discussing the draft convention on the legal status of the aircraft commander. In Chapter V [8] we shall deal with the extent to which this responsi-bility of the aircraft commander is limited through the obli-gation to comply with instructions "from the ground."

The existing Dutch legislation does not contain such a wide definition of the aircraft commander's responsibility. Admittedly the ban on performing "air navigation in such a way that public order or public safety is disturbed or endangered" naturally applies to the aircraft commander in the first place. But this provision of the Aviation Act [9] only refers to violations of *public* order or safety, i.e. proceedings which are not a matter of private concern. [10]

Although there is no general statement of the aircraft comman-der's responsibility for safe execution of the flight, it is apparent

---

[1] Act of 25th March 1936, article 11. cf RFDA 1950, p. 100.
[2] Maschino, "La Condition Juridique du Personnel Aérien" pages 128 and 134.
[3] Annex 6 par. 4.3.2.
[4] Annex 6 par. 4.3.3.
[5] Annex 6 par. 4.3.4.
[6] Annex 6 par. 4.5.1.
[7] See page 117.
[8] See page 50.
[9] Aviation Act, article 10.
[10] Decision of the High Court of Justice of 29th June 1936, N.J. 1001; Cf. Goedhuis, "Handboek voor het Luchtrecht" page 169.

from a large number of special provisions that the Dutch legislator holds the aircraft commander directly responsible again and again for correct observance of the most diverse regulations. [1]

Thorough knowledge of these regulations is naturally of fundamental importance to him. [2]

As most of the regulations in question are either technical or administrative, for the purpose of this study it will be sufficient to list some of the subjects in order to furnish an insight into the number and nature of the rules and regulations which the aircraft commander must obey under penalty of sanctions.

Aviation Act

Prohibitions against flying an aircraft with military markings [3], flying over forbidden areas [4], flying without a valid certificate of competency [5] and certificate of airworthiness [6], and landing on parts of the territory of the Netherlands not designated as aerodromes [7]; obligation to show documents. [8]

Regulations for State Control of Air Navigation (R.T.L.)

Prohibitions against flying without valid markings [9]; instructions concerning composition of crew [10]; prohibition against flying if no commander is designated [11]; prohibition against aerobatics [12]; instructions concerning repairs [13]; signature of daily inspection certificate [14]; prohibition against carrying parachutes [15]; prohibition against taking off without the permission of the airport manager [16]; compliance with instructions of the airport manager [17];

---

[1] For that matter just as the ICAO, see in particular Annex 2, 6 and 9.
[2] A practical textbook used for the training of Dutch transport pilots is Dubois, "Luchtvaartwetten en Verkeersvoorschriften".
[3] Article 6.3.
[4] Article 8.3.
[5] Article 15.1.
[6] Article 15.2.
[7] Article 42.
[8] Article 50.
[9] Article 18.
[10] Article 67, 68, 69 and 70.4.
[11] Article 70.3 .
[12] Article 77.2.
[13] Article 86.3.
[14] Article 87.6 and 87.8.
[15] Article 145.
[16] Article 169.
[17] Article 170.

carrying and completion of flight documents [1]; prohibition against using a non-transport plane for transport purposes [2], admitting unauthorized persons to the cockpit [3], carrying drunken or dangerous passengers [4], carrying explosives or inflammable substances [5], carrying sick or dangerous animals [6] and carrying passengers on test flights [7].

Air Traffic Regulations

The commander is responsible for compliance with all the air traffic regulations [8], which include sections dealing with: Lights and daylight markings, signals, general provisions concerning air traffic, special provisions concerning traffic on or in the vicinity of airfields, provisions of a general nature.

The aircraft commander is also directly concerned with the provisions of the *Act for the Regulation of the Sanitary Control of Aerial Navigation* [9] and the Decree [10] giving effect to that Act, as well as the *Customs Decree* [11].

In this brief summary we wish to confine ourselves to the statutory regulations which the commander has to obey, otherwise we should have to list the official administrative orders such as the *Notices to Airmen* (published by the Netherlands Department of Civil Aviation) and also *Company regulations.*

A more important point is that when flying over the territory of other States the commander of a Dutch aircraft must adhere to the local instructions in force concerning flight operations. This follows from Art. 12 of the Chicago Convention, which reads as follows:

"Each contracting State undertakes to adopt measures to insure that every aircraft flying over or manoeuvring within its territory and that every aircraft carrying its nationality mark, wherever such aircraft may be, shall comply with the rules and

---

[1] Article 184, 184 ter, 184 quater, 185, 185 bis.
[2] Article 189.
[3] Article 193.
[4] Article 195.
[5] Article 196.
[6] Article 196.4.
[7] Article 196 septies.
[8] Article 2.
[9] Act of 26th October 1935, S. 626.
[10] Decree of 11th February 1936, S.840.
[11] Decree of 27th December 1928, S.500.

regulations relating to the flight and manoeuvre of aircraft there in force . . .".

The operator is obliged to see to it "that his employees when abroad know that they must comply with the laws, regulations and procedures of the States in which his aircraft are operated." [1]

Fortunately the countries which are members of ICAO have agreed to keep their regulations uniform "to the greatest possible extent" with those established under the Convention. [2]

As departures from the Standards and Recommended Practices formulated by ICAO are permissible, however, and as ICAO cannot make, or at least has not yet made, regulations covering every field of air navigation, there are still a great many national and regional differences. It is obvious that this greatly increases the difficulty of the aircraft commander's task.

The present Dutch Aviation Act only applies "on and above the surface of the earth within the territory of the Netherlands," [3] but the intention is that the new Aviation Act shall apply to Dutch aircraft wherever they may be. The commander of a Dutch aircraft outside Dutch territory will then be simultaneously subject to two sets of regulations, viz. the Dutch regulations and those of the foreign country in question.

In our opinion it may be concluded from the foregoing that in the performance of his duties the aircraft commander is subject to a regimentation which is not only very extensive and detailed but also extremely unclear.

One pilot was undoubtedly exaggerating when he remarked: "Instead of a co-pilot, I need a lawyer in the cockpit to tell me what to do. I don't think first what is the safest thing to do, but what is the legal thing to do. If I break rule so-and-so, will I be grounded and fined?" [4] And yet there is a lot of truth in the observation made by a representative of IATA at an ICAO Meeting that "multiple regulations, each of which in itself might be completely logical and reasonable, collectively proved to be confusing and fatiguing to the pilot". [5]

---

[1] Annex 6, par. 3.1.
[2] See page 12.
[3] Aviation Act article 1, sub I, cf. Spanjaard, "Vliegwereld" 6th April '50, page 230.
[4] Nicholson, "Air Transportation Management" page 129.
[5] At the 4th Meeting of the Operations Division, March-April, 1951, Appendix A to Doc. 7151 — OPS/609, page 144.

CHAPTER V

INSTRUCTIONS FROM THE GROUND

The technical progress of aviation is accompanied by the multiplication of new and more highly specialized functions. This tendency is perceptible in the composition of the crew. At first the pilot — who was often the constructor of the aircraft as well [1] — flew entirely alone, but in course of time the crew had to be expanded, part of the pilot's work being entrusted to a co-pilot, a navigator, a flight engineer and a radio operator.

As the latter crew members perform their duties under the supervision and on the responsibility of the aircraft commander, so that it is purely a question of delegation of work, this change has in no way affected the position and responsibility of the commander. One might say that the frequently individualistic "pilote-aviateur" has now attained the status of leader of a team of specialists.

It should be noted that the latest technical developments have reversed the tendency towards expansion of the crew. Through higher speeds and consequently shorter flights, there is less necessity for a large crew, while the gradual perfecting and automatic control of the mechanical aids on board make it less essential that these should be operated by specialists.

But we must now turn our attention to another development, viz. the development whereby certain powers and responsibilities of the aircraft commander are transferred to authorities on the ground, and whereby the commander is obliged to obey instructions from the ground.

This state of affairs originated from the improvements in communications, viz. radio telegraphy and radio telephony, and it was primarily due to the growing volume of air traffic, coupled with the increasing amount of flying in cloud and at night. In

---

[1] For example the brothers Wright, Blériot, Farman, Fokker.

order to reduce the resultant danger of collision, it became necessary to control air traffic from the ground; the organizations created for this purpose were designated as "Air Traffic Control" (A.T.C.).

A more recent innovation arose from the desire of the airlines to be able to give instructions direct to their aircraft commanders during the flight, with a view to operating their fleets as safely and as profitably as possible. This is known as "Operational Control."

With regard to this development, Le Goff remarks:

"We should note that technical progress, wireless control particularly control of aircraft from the ground by specialized services, have transformed completely the responsibility of the carrier. Problems were different before, the pilot on board of the aircraft was considered absolutely independent, to-day his role on board is reduced. As the aircraft is directed from the ground, the pilot follows only orders given him from the ground. Thus arises the responsibility of public services, which give the orders." [1]

There are various drawbacks about dealing with this subject in the present study. On the one hand there is the preponderantly technical nature of the subject, and on the other hand it must be remembered that the aviation world's views on the subject are by no means stablilized. [2] Since the development in question may affect the status of the aircraft commander to a considerable extent, in our opinion it is essential to examine the whole position. In this respect we share the view expressed by Lemoine, when commenting on the technical regulations concerned: "La connaissance de cette réglementation n'est pas inutile pour le juriste car il arrivera qu'elle constitue un élément de décision important dans des questions de responsabilité. Elle peut aussi avoir un intérêt direct en matière d'assurance." [3]

AIR TRAFFIC CONTROL

A wide network of services, collectively designated as "Air Traffic Services", has been established on behalf of air navigation in all parts of the globe.

---

[1] Le Goff, "The Present State of Air Law" page 24.
[2] Cf. Bucher, "Le Statut juridique du Personnel Navigant de l'Aéronautique", page 76.
[3] Lemoine, "Traité de Droit Aérien" page 272.

The States which are members of ICAO must indicate the parts of the air space above their territory for which such services will be provided; they must then institute these services and supervise them, although the latter responsibility may be delegated to other States [1]. The Air Traffic Services comprise the Air Traffic Control Service, the Flight Information Service and the Alerting Service. We shall confine ourselves to a brief discussion of the task and function of the Air Traffic Control Service, which is of most importance for our subject.

The task of A.T.C. is to promote a "safe, orderly and expeditious flow of air traffic" [2]. As elaborated in Annex 11 [3] this means that A.T.C. must:

"1. prevent collisions between aircraft;
2. prevent collisions on the manoeuvring area between aircraft and obstructions;
3. expedite and maintain an orderly flow of air traffic".

A.T.C. service is only furnished in controlled air space, and exclusively for flights which take place under Instrument Flight Rules; in the vicinity of airfields, however, all flights are controlled [4].

It must be explained that there are two categories of air traffic rules, viz. Instrument Flight Rules (IFR) and Visual Flight Rules (VFR). The first set of rules apply in control areas if the visibility is less than 5 kilometres and the distance from the aircraft to the clouds less than 150 metres vertically and less than 600 metres horizontally [5]. The commander is responsible for deciding whether a flight can take place under IFR or VFR [6].

It is apparent from the foregoing that the task of A.T.C. is limited in the following respects:

a. Spatially, because it only functions in designated, limited areas. Large areas are either completely uncontrolled, or else only controlled above or below a certain altitude.

b. Depending on weather conditions, since it generally functions

---

[1] Annex 11, par. 2.1; for the Netherlands control area this service is provided by the Air Traffic Safety Division of the Department of Civil Aviation.
[2] Annex 2, Chapter 1, Definitions.
[3] Annex 11, par. 2.2. and par. 2.3.
[4] Annex 11, par. 3.1.1.
[5] Annex 2, par. 2.3.
[6] Annex 2, par. 2.3.4.

only for flights under IFR. But even in VFR weather conditions the commander can voluntarily place himself under the control of an A.T.C. unit by stating that he wishes to fly under IFR and applying for air traffic clearance. In that case, however, he remains personally responsible for avoiding danger of collision. [1]

c. A final important limitation is that A.T.C. is not called upon to prevent aircraft from colliding with fixed objects like mountains and other obstacles. This limitation is especially important because such collisions are a relatively frequent cause of accidents.

A brief comment on the liability of A.T.C. may be appropriate at this point. Under Dutch maritime law the State is not liable for acts and omissions of a ship's pilot [2]. Although the task of A.T.C. is somewhat the same as the task of a ship's pilot, e.g. when an aircraft is brought in to land by means of bearings, in our opinion this rule cannot be applied by analogy in the absence of a similar provision in air law. We hold that in certain cases the State can be held liable for damage resulting from incorrect instructions from A.T.C., under the terms of Art. 1401 of the Netherlands Civil Code. It should be borne in mind, however, that the court must take into account the limits within which the authorities are free to act at their own discretion [3]. The court's interpretation will ultimately determine whether a certain incorrect instruction came within the field reserved to the authorities (to our mind, the plea that only "advice" is given does not hold good, on account of the wording of the ICAO regulations); Dutch judiciary law is still lacking on this point.

In American judiciary law there are numerous cases arising out of supposed shortcomings in services performed by A.T.C. or allied authorities. As A.T.C. is a function performed by or on behalf of the authorities, one is confronted with the principle of "sovereign immunity" under Anglo-Saxon law, whereby an

---

[1] Doc. 4444-RAC/501 par. 2.1.5.1.4. under (1). The wording and numbering, but not the substance, of this document have recently been revised (Doc.4444-RAC/501/1).

[2] Cleveringa, "Het Nieuwe Zeerecht" page 237.

[3] Decision of the High Court of Justice of 29th March, 1940, N.J. 1940, No. 1128; see for detailed account Honig, „Overheidsaansprakelijkheid en luchtvaart", NJB 1951, page 767.

action can only be brought against the authorities with their consent.

Eastman [1] describes this principle as archaic and expresses the view that the rules of common law ought to apply in regard to liability. The liability would then be based on the principle of a "gratuitous undertaking," in which case a person is liable for "failure to exercise with reasonable care such competence and skill as he possesses or leads the other reasonably to believe that he possesses". On the other hand, no liability would be incurred by discontinuing such services if the other party is not placed in a less favourable position than that prevailing when the services were started.

How does all this affect the position of the aircraft commander? The following is laid down as a guiding principle in the "Rules of the Air": "The pilot in command of an aircraft shall be directly responsible for the operation of the aircraft and shall have final authority as to disposition of the aircraft while he is in command." [2] But immediately after this comes the provision that the aircraft commander "shall be responsible for compliance with air traffic control instructions received." [3]

It appears difficult to reconcile these two statements and one cannot help feeling that those responsible for drafting the rules have halted between two ideas.

In the first clause we have the principle of the aircraft commander's responsibility and his prerogative of conducting the flight in accordance with his own judgment, while in the second clause we are faced with the necessity of making the A.T.C. instructions mandatory for the commander. If the aircraft commander were not obliged to comply with such instructions, it would be difficult for A.T.C. to accept responsibility for prevention of collisions. With due observance of the first principle, we consider that the latter ought to prevail, this being in the interests of the aircraft commander too.

In order to eliminate the contradiction between the two clauses, the phrase "subject to his duty to comply with A.T.C. in-

---

[1] Samuel Ewer Eastman, "Liability of the Ground Control Operator for negligence", JAL 1950, page 170.
[2] Annex 2, par. 3.1.1.
[3] Annex 2, par. 3.1.2.

structions" (or words to that effect) should be added to Par. 3.1.1.

At the same time the aircraft commander must be allowed an opportunity to deviate from the A.T.C. instructions under certain circumstances. This is provided for in the "Procedures for Air Navigation Services, Air Traffic Control," [1] but only in cases of emergency.

In that event A.T.C. must be informed of the deviation as quickly as possible. A similar provision ought to be included in Par. 3.1.2. of the "Rules of the Air" but with its scope extended to embrace "emergency or other unforeseen circumstances."

As already remarked, under VFR conditions the aircraft commander is directly responsible for avoiding collisions with other aircraft, even if the flight is being carried out in accordance with A.T.C. instructions. [2] In this respect "the information and instructions issued by aerodrome control towers are merely intended to aid pilots in-command to the fullest extent". [3] In other words, the "instructions" are not binding in this case. The purpose of the above regulation is to oblige the aircraft commander to keep a good look-out himself, if this is possible under the prevailing weather conditions.

In conclusion it must be pointed out that the granting of an air traffic clearance does not entitle the commander to disregard any instruction issued by the competent authorities in connection with air safety [4]. Although the reasoning behind this regulation is perfectly clear, in our opinion it is superfluous.

After all, an air traffic clearance only means that A.T.C. has authorized performance of a flight on certain conditions (particularly in regard to course, speed, altitude, etc.) [5]. A clearance of this nature is only issued with a view to traffic control. Obviously such an authorization does not in any way relieve the commander of his obligation to comply with instructions, just as a green traffic signal does not mean that it is permissible for a road user to drive a vehicle with defective brakes or without a driving licence. Accordingly it is not up to the A.T.C. personnel to ascertain whether the safety regulations are being complied with before

---

[1] Doc. 4444, par. 2.1.5.1.4. under (3).
[2] Doc. 4444, par. 2.1.5.1.4. under (1).
[3] Annex 11, note under par. 3.1.1.
[4] Doc. 4444, par. 2.1.5.1.4. under (2).
[5] Annex 2, Chapter 1, Definitions.

granting a clearance. After an accident to a Dutch aircraft, the Netherlands Air Accident Board replied in the negative when a question was raised as to "whether the Air Traffic Control personnel at Schiphol ought not to have granted a traffic clearance for the flight in view of the fact that the aircraft's equipment did not satisfy the relevant regulations" [1].

OPERATIONAL CONTROL

Operational Control is of a nature entirely different from Air Traffic Control. The latter is a function exercised by or on behalf of the authorities, whereas the former is a system of control and management during the flight, developed and exercised by the airlines themselves.

At the Third Session of the Operations Division of ICAO, attention was drawn to the desirability of arriving at a clear distinction between the task of A.T.C. and the task of the operator in regard to the conduct of a flight. The original, extremely wide description of the task of A.T.C., as stated in Annex 2 ("to promote the safe, orderly and expeditious flow of air traffic") made sharper delimitation advisable.

The airlines were particularly anxious that their power to give their aircraft instructions and directions during the flight, in the interests of safety and economy, should be explicitly defined. In view of this it was recommended that the task of A.T.C. should be more clearly defined (this has since been done in Annex 11) [2].

At the same time the following resolution was adopted with regard to the definition of "operational control" and the allocation of responsibility for such control [3].

"a. *Definition of operational control.*

The exercise of authority over the initiation, continuation, diversion or termination of a flight.

"b. *Standard allocation of responsibility for operational control.*

An operator or his designated representative shall have responsibility for operational control.

"*Note.* The above does not infringe upon the rights of a

---

[1] Sentence of the Aircraft Accident Board dated 2 nd January 1950.
[2] See page 52.
[3] Final Report of the Operations Division, 3rd Session, Doc. 6640, page 175.

State in respect to the operation of aircraft registered within that State".

Although this did not furnish an entirely satisfactory delimitation of the responsibility of A.T.C. and Operational Control, the foregoing may be regarded as an important step in the right direction. It must be borne in mind that the form and contents of the two services are not yet stabilized, and for the time being it is therefore difficult to reach international agreement on a clearer definition. Accordingly, we cannot share the view expressed in a reservation made by the French delegation that "the concept of operational control, which is a source of confusion, should be eliminated" [1].

On the other hand one can agree with the objection that the resolution does not in any way clarify the division of responsibility between the aircraft commander and the authority exercising the operational control.

Before going into this in more detail, the method adopted for operational control in actual practice will be briefly explained.

According to Annex 6, an operator is obliged to set up an organization to supervise flight operations. The method of supervision is subject to the approval of the State where the aircraft are registered [2].

In the United States, and later in other countries as well, many carriers have built up organizations based on the stationing of "Flight Operations Officers" at the airfields regularly used by their aircraft.

These officials may be regarded as the operator's "designated representatives" (within the meaning of the definition contained in Annex 11) for purposes of operational control. They have to meet high requirements in regard to their knowledge of navigation, meteorology, rules and regulations, etc. [3], and they are often recruited from former flying personnel. In order to familiarize themselves with the routes under their supervision, they must make regular flights on these stretches. [4] The duties of the Flight Operations Officers (who are also referred to as "dispatchers") [5]

---

[1] Doc. 6640, page 190.
[2] Annex 6, par. 4.2.
[3] Cf. recommendation in Annex 1, par. 4.3.
[4] Annex 6, par. 10.2.
[5] In the Netherlands the name "flight adviser" has been officially accepted; by

include helping the aircraft commander with the flight preparations by giving him all the information that he may require; they assist the aircraft commander in drawing up the flight plan, which the aircraft commander must sign for agreement; while the flight is in progress they must remain on duty (keep a "flight watch") and supply the commander with the information needed for a safe flight; in cases of emergency they must take the steps prescribed in the Operations Manual. [1] In performing his duties the Flight Operations Officer must take care not to come into conflict with the procedures of A.T.C., the Meteorological Service and the Communications Service. [2] Good coordination must be secured [3].

In the preceding paragraph we have described the duties as specified in sundry ICAO regulations. In actual practice, [4] however, there is a perceptible tendency to give the Flight Operations Officer more extensive powers. The relationship between the Flight Operations Officer and the aircraft commander is frequently depicted as one of joint responsibility. Before the commencement of a flight, the Flight Operations Officer and the aircraft commander must reach agreement on the execution of the flight, as regards the route to be flown, the loading of the cargo, etc. The Flight Operations Officer can instruct the aircraft commander to alter his route during the flight. Such instructions are not absolutely binding, however, in that the commander is entitled to deviate from them — though he must afterwards submit a reasoned report to the management of his company.

As long as the idea of joint responsibility and the power to give instructions to the captain are not confirmed by law, in our opinion internal arrangements of this nature cannot limit his responsibility.

Moreover, the question must be examined whether the development of the commander's responsibility along the lines indicated above is desirable.

resolution of the "Minister van Verkeer en Waterstaat" of 26th November 1951 a committee has been instituted for examination of "flight advisers".

[1] Annex 6, par. 4.6.
[2] Annex 6, par. 4.6.2.
[3] Doc. 4444, page 8a.
[4] For the activities of the "dispatcher" in the U.S.A. see: Baker, "Airline Traffic and Operations" page 165; Frederick, "Commercial Air Transportation" page 601; Zweng, "Airline Transport Pilot Rating" page 433; Bullock, "Airline Piloting" page 160; Speas, "Airline Operation" page 27.

LANDINGS UNDER UNFAVOURABLE WEATHER CONDITIONS

The question of the aircraft commander being obliged to obey instructions from the ground arises in its most pressing form in cases where unfavourable weather conditions make it necessary to decide whether the aircraft can land or not. If the answer to this question is in the affirmative, the second question to be raised is whether the power to issue such instructions should rest with A.T.C. or the authority exercising operational control, i.e. the Flight Operations Officer on duty.

In aviation circles there is some controversy over the answers to these questions. Economic, psychological and safety considerations all play a part. It is important that the point should be settled because accident statistics show that a large percentage of accidents to commercial aircraft occur during attempts to land under unfavourable weather conditions [1].

In an article published in „The Aeroplane" [2] the author discusses fifteen recent air accidents in which 49 crew members and 130 passengers lost their lives while 63 persons were injured. He concludes that "a careful study of the accidents listed reveals a common factor in that they were all preventable, and not only that, they were preventable by action instituted from the ground." Furthermore "... the evidence presented is convincing enough for the argument, that the responsibility for a decision (to land or divert) should not rest solely with the captain".

Another writer [3] asserts that " ... of the aircraft accidents involving scheduled services for the past two years, the majority could have been prevented by positive action from someone on the ground either by diversion, holding or rerouteing".

Still another opinion [4] to support the view that a landing under unfavourable weather conditions should be forbidden, reads as follows: " ... if the airfield is left open, the pilot, I fear will always be human enough to cite emergency and prove his skill; whereas rigorous discipline and no excuse may send him to another airfield and save a hundred lives a year".

---

[1] According to data of ICAO approximately 30% of the accidents take place in landing; see Aircraft Accident Digest No. 1, Circular 18-AN/15, page 9.

[2] "The Case for Flight Despatch", The Aeroplane, July 20th 1951, page 77.

[3] R. D. Cooling, "Aircraft despatch and Flight Supervision", Aeronautics, August, 1950, page 24.

[4] R. L. Hanson, Flight, 16th November, 1950, page 444.

A special argument [1] for stricter control from the ground applies to jet aircraft, whose particular characteristics make it probable that " ... the pilot of jet aircraft will have to rely to a much greater degree on the advice and judgment of the ground staff as to whether he should decide to divert".

On the other hand, there is an understandable aversion — in the first place among pilots — to accept a situation whereby the aircraft commander's decisions would be subordinated to directions or instructions from the ground, especially if these were to emanate from frequently anonymous personnel whose status, responsibility and qualifications are still uncertain. In view of this we consider it somewhat unfair to say that the pilots "are fighting to retain the omnipotent position that they enjoyed in so vastly differing circumstances prior to the War, much in the same way that the Stuarts fought for the divine right of kings". [2]

One of the exponents of the above-mentioned "pilot's view" is Air Vice-Marshal D. C. T. Bennett, [3] who is " ... strongly in favour of reducing control from the ground and giving the aircraft captain a greater measure of responsibility."

Members of the American Airline Pilots Association are reported to have " ... damned GCA [4] from the first as an instrument for taking authority out of the cockpit and vesting it in someone on the ground" [5].

At its fourth conference in Copenhagen the International Federation of Airline Pilots Associations adopted a resolution that " ... assessment of weather conditions as made by the pilot shall always be accepted as final".

Even Jerome Lederer [6], President of the Flight Safety Foundation, holds that "we have too many people on the ground who are confused about their position in respect to authority. There can only be one person commanding an aircraft ... So let's keep

[1] John Longhurst, "Jets and Air Traffic Control", The Aeroplane, August 18th, 1950, page 195.
[2] The Aeroplane, July 20th, 1951, page 77.
[3] The Aeroplane, February 24th, 1950, page 203.
[4] GCA: "Ground Controlled Approach" a landing system whereby the flight of an aircraft is followed on the ground by means of radar, and radiotelephonic instructions are given to the captain as to how the aeroplane must be flown to ensure a safe landing, so-called "talk down".
[5] Aviation Age, November 1950, page 17.
[6] Flight Safety Foundation, Accident Prevention Bulletin 50–18.

command in the cockpit, let the ground give information and remember, they are not in that airplane, they don't see all the factors the pilot sees".

Moreover, several writers specialized in air law share this point of view.

In an article on "The Aircraft Commander in International law" Knauth presents the following view: ,,No less qualified man should ever be able to tell the more qualified first pilot what to do in any emergency. The decision to turn back, to go to an alternate airport, to land or not to land in thick weather, etc. must rest with one man — the most highly trained man aboard, up-to-the minute in his knowledge of the procedures and practices of the place and the hour. No employer, company officer or stockholder should be able to overrule his judgment" [1].

Lemoine has made the following remark on the subject: ,,En vol, les appareils modernes restent en contact avec les services au sol qui leur fournissent toutes les indications nécessaires à une heureuse navigation. Mais il faut bien se rendre compte que les relations qui s'établissent ainsi ne doivent pas porter atteinte à la libre décision du commandant." [2]

In conclusion it is worth mentioning the authoritative voice of a British Court of Inquiry which, although it attributed an accident to a commander's endeavour to land under weather conditions "in which it was imprudent for him to do so," nevertheless concluded: " ... I do not think it necessary to recommend any change in the existing system by which the pilot is ultimately responsible for the decision whether or not to divert. Indeed, I think that if the decision whether or not to land was taken from him, the pilot might be encouraged to attempt a landing in all cases in which he was not instructed from the ground to divert." [3]

WEATHER MINIMA

The "weather minima" form a standard for judging if the weather conditions are such that a landing can safely be effected. Weather minima may be defined as the minimum values of

---

[1] JAL 1947, page 161.
[2] Lemoine, ,,Traite de Droit Aérien" page 237.
[3] Report of the Court Investigation on the accident to Viking G-AHPN on 31st October, 1950.

visibility and ceiling height below which a landing must not be attempted at a particular aerodrome [1]. The process of establishing such minima is extremely complicated, since factors such as the skill and experience of the pilots, the presence of obstacles in the vicinity of the airfield, the characteristics of the type of aircraft used, the nature of the landing aids, etc., must all be taken into account. [2]

In principle, the fixing of weather minima might be regarded as a duty of the State in which the airfield is situated, the State where the aircraft is registered, or the airline concerned (with or without the approval of the State of Registry).

ICAO does not take up any definite standpoint in this connection [3]. It is prescribed that an airline must lay down weather minima for every aerodrome used, but these minima must not be lower than those that may have been established by the State in which the airfield is situated; it is expressly added that the latter State is not obliged to establish weather minima.

Similar regulations apply to non-scheduled operations, on the understanding that it will suffice if the airline specifies the method by which the weather minima shall be determined; this method, however, must be approved by the State in which the aircraft is registered. In connection with the above, the following regulation contained in Annex 6 is also of importance:

"During their published hours of operations and subject to their published conditions of use, aerodromes and their facilities shall be kept continuously available for flight operations irrespective of weather conditions" [4].

From a closer study of this regulation it follows that there is nothing to prevent the aerodrome authorities from laying down weather minima and forbidding a landing if the weather conditions are lower than these minima. In such a case the aerodrome weather minima may be regarded as a "published condition of use".

---

[1] Instead of in weather minima the limits can also be expressed in the so-called "critical hight" which is defined as "the height above aerodrome elevation at which descent during instrument approach should be discontinued if the approach cannot be continued visually" (Report of the Operations Division, 4th Session, Doc. 7151–OPS/609, page 51).

[2] See Annex 6, par. 4.2.5.

[3] ibidem.

[4] Annex 6, par. 4.1.3.

All this has led to great diversity in the practice adopted in different countries [1].

In some countries (e.g. Denmark, Sweden and the U.S.A.) the authorities establish weather minima for every airfield, and both national and foreign aircraft must adhere to these minima.

In other countries (e.g. the Netherlands and the United Kingdom) the fixing of weather minima is considered to be entirely the responsibility of the airline concerned with or without the approval of the State where the aircraft are registered.

Schiphol Airport (Amsterdam), for example, remains open to visiting aircraft even under the most unfavourable weather conditions.

A Committee under the chairmanship of Lord Brabazon of Tara was set up by the British Minister of Civil Aviation to inquire into "the relative responsibilities of the captain of an aircraft, the operator and the aerodrome authority in deciding whether an aircraft can safely land at, or take off from, an aerodrome in bad weather conditions".

This Committee came to the conclusion that it is not advisable for the State to establish such weather minima, partly in view of the following: [2]

" ... a proposal has been made that it would be reasonable for the State to close aerodromes at a weather limit below which any attempt to land or to take off might well be disastrous. Such a system, of course, still envisages that operators would lay down weather minima above these limits. Although the view was put to me that there need be no danger of the weather limits at which the State closes the aerodrome being regarded as the effective weather minima, it appears to me that there would most certainly be a tendency amongst pilots to take the view that when an aerodrome was open it was safe to attempt to land. This fact is of paramount importance, for although a minority of accidents would have been prevented by closing aerodromes at the low limits envisaged in the above proposal, a larger proportion have, in fact, occurred in conditions above these values. In the end,

---

[1] Cf. Lord Brabazon of Tara, "Landing and Taking-off of Aircraft in Bad Weather", Command paper 8147.

[2] "Landing and Taking-off of Aircraft in Bad Weather" pages 11 and 12.

therefore, adoption of the proposal would, in my opinion, result in an increase, rather than a decrease, in the number of accidents . . .".

On the other hand, the Committee considered that:

" . . . it would be wrong to leave the decision entirely to the pilot. Not only would it place an unwarrantable burden upon him but it would also initiate a practice already proven dangerous. For instance, there is evidence that take-off and landing accidents have occurred which have revealed pilot error as a factor. I should like to say at once that I do not believe that any trained airline pilot consciously compromises safety but I do consider that on psychological grounds, a number of pilots are influenced into attempting a landing in dangerous conditions. There would accordingly appear to be a need for some overriding control and I am satisfied that this should remain an operator's responsibility . . .".

There is little disagreement on the principle that the aircraft commander should be tied down to certain standards in deciding whether he can land or not — though there are different views as to who should fix these standards — but it is a different matter to decide who is ultimately responsible for judging whether the weather conditions are below the minima or not. Meteorological observations of the ceiling height, amount of cloud and visibility may vary greatly, depending on the time and place. As long as objective observation (e.g. by electronic means) is not generally applied, the observations will also be subjective, i.e. dependent on the observer's estimate.

It follows from this that the ceiling and visibility reported by the meteorological service will frequently differ from the pilot's estimate. It must be remembered that the aircraft commander assesses the visibility differently, viz. from above and at an angle (as well as from a different place, since the meteorological station is generally on the airfield); rain or ice forming on the cockpit windows may adversely affect his estimate, while the brightness of the runway or approach lighting may have a favourable influence on it. Moreover, the commander will often be forced to decide very quickly whether to land or to abandon the attempt and climb to a higher altitude; it is quite possible that under such circumstances he will judge the situation as a whole and not compare it with a ceiling and visibility estimate expressed in numerical values.

The intention of ICAO is to forbid landings below the weather minima [1], and in certain countries sanctions can be imposed for failure to adhere to such minima. For the reasons already mentioned, however, it is extremely difficult to enforce prohibitions of this nature.

Naturally the safest system would be to forbid a landing if the meteorological service has established and reported that the weather conditions are below the minima, even when the pilot's own observations lead him to believe that this estimate is too pessimistic. Furthermore, a landing would also have to be forbidden if the pilot himself observed that the weather conditions were below the minima; in that case he should be unable to plead that an over-optimistic estimate had been given from the ground.

There are objections to this system, of course, in cases where it is doubtful whether the local meteorological service functions properly. A landing might then be forbidden unnecessarily, with a serious financial loss as a result. IATA therefore adopts the standpoint that the aircraft commander must have the final say in deciding whether a landing is possible, with due observance of the weather minima fixed for the airfield in question. This view finds support in the opinion expressed by ICAO that "the actual visibility conditions from the pilot's cockpit ... cannot normally be adequately assessed by existing meteorological techniques from the ground, but must be assessed by the pilot in the aircraft." [2]

The British Committee mentioned earlier gave a negative answer when asked if A.T.C. should be given the power to issue binding instructions on whether to land or not. As regards granting such a power to the Flight Operations Officer, the Committee refrained from pronouncing on this subject because its implications would extend beyond the Committee's terms of reference. [3]

It is undoubtedly correct to say that a number of accidents might have been prevented by granting powers of this nature to

---

[1] See Annex 6, par. 4.4.1.
[2] Report of the Operations Division, 3rd Session, Doc. 6640-OPS 567, page 46.
[3] loc. cit. page 13.

an official on the ground. The opposing argument is that faulty decisions can also be taken on the ground, particularly because the people on the ground are not able to assess and weigh up the circumstances of the flight in the same way as the aircraft commander does during the flight. Another point is that the granting of such far-reaching powers to officials on the ground may give rise to uncertainty about command relationships, which can be dangerous in itself. It has therefore been rightly remarked that "increased authority to the controllers is a double edged weapon, for its limits might not be easy to define."

In the foregoing we have merely outlined some of the problems and controversies, without attempting to offer a solution. A practical solution will have to be developed ultimately, though perhaps the question will lose its urgency through the introduction of landing aids which will enable aircraft to land under all weather conditions [1].

INSTRUCTIONS FROM THE AIRPORT MANAGER

In addition to instructions from A.T.C. and Operational Control, in the execution of his duties the aircraft commander is also concerned with the instructions of the airport authorities, and the airport manager in particular.

Dutch air legislation contains two provisions on this subject, though it is more than twenty years since they were enacted. In the first place, it is forbidden for the commander of an aircraft to take off from an aerodrome without having been given permission to do so by or on behalf of the airport manager [2]. In actual practice the permission to depart is given by the control tower, i.e. by A.T.C., which implies tacit consent of the airport manager. By means of this regulation, however, the airport manager can prevent an aircraft from taking off if the landing fees have not been paid, if the crew have committed an offence, or if the aircraft's papers are not in order.

In the second place, the commander of an aircraft on or above an aerodrome is obliged to comply with the instructions given to

---

[1] Cf. "All-weather Era here, say experts, "Aviation Week, 9th July, 1951; "All-Weather Approaches", Flight, 13th July, 1951, page 52.
[2] Article 169 RTL.

him by or on behalf of the airport manager with a view to safety and order [1]. As far as the importance of this regulation is concerned, it must be remarked that air traffic is normally controlled by A.T.C. nowadays and rarely by the airport manager. [2]

Although airfield organization and the associated services may assume many diverse forms, [3] American development in this respect can serve for purposes of illustration.

Originally the air traffic on and above an aerodrome was controlled by the airport manager, usually with the aid of visual signals because radio communication facilities were seldom available. When congestion in the vicinity of airfields began to form a source of danger, the airlines co-operated with the airport authorities in setting up a simple form of air traffic control.

As far as traffic on the airways is concerned, control was taken over by the federal government in 1936, while the control arrangements on and around airfields were left to the local authorities. During the last World War, the control of all air traffic was entrusted to the C.A.A. and this arrangement was continued after the end of the war. [4] In the Netherlands the Air Traffic Safety Division of the Netherlands Department of Civil Aviation is responsible for the control of air traffic.

Nowadays the relationship between the airport authorities and A.T.C. is often such that the airport manager tells the control tower which runway can be used for landing, which taxiway is available, etc [5]. The control tower regulates the traffic on the basis of this information, maintaining direct contact with the aircraft and giving instructions if necessary. In other words, the airport manager himself does not have any contact with the aircraft.

Although the traffic is not actually controlled by the airport manager any longer, the above mentioned provision of the R.T.L. has more than historic interest.

The airport manager can, and does, issue instructions about the

---

[1] Article 170 RTL.
[2] See page 51.
[3] For the various forms of organization see Zweng, "Airport Operation and Management"; Bollinger, "Terminal Airport Financing and Management"; Frederick, "Airport Management".
[4] Cf. Frederick, op. cit. page 154.
[5] Cf. C.A.A. Journal, Vol II, August 15th 1950, No. 8, page 86.

parking of aircraft, about anchoring them when there is a strong wind, about the method of refuelling, etc.

In all these cases the aircraft commander is obliged to comply with the instructions given to him by or on behalf of the airport manager.

CHAPTER VI

SEARCH AND RESCUE

At the present stage of technical development, a forced landing on the sea or outside an aerodrome is a very rare occurrence. It has been calculated [1] that only one ditching need be anticipated in every 33,000 North Atlantic crossings. On the one hand this figure is too pessimistic because it is based on pre-war statistics relating to engine failure, and on the other hand it is too optimistic because no allowance is made for such factors as shortage of fuel through excessive icing, inaccurate navigation or unexpectedly strong headwinds, but it has been found that such occurrences are indeed very rare.

The technical regulations drawn up by ICAO are primarily intended to prevent accidents of this type. Nevertheless, should a forced landing be necessary, there are other ICAO regulations whose object it is to minimize the disastrous consequences of such an occurrence and to afford the maximum chance of rescuing the survivors.

It is prescribed, for example, that transport aircraft must carry certain emergency equipment on flights over sea, e.g. life belts, life rafts and signalling kit [2], while the crews must be regularly trained and tested in the use of the equipment and in the procedure to be followed in the event of a forced landing [3]. The operator must also give the passengers full information concerning the location and use of the available emergency equipment. [4]

These regulations ought to be supplemented by an internationally accepted obligation, on humanitarian as well as utilitarian grounds, whereby assistance must be rendered by all who are in

---

[1] McFarland, "Human Factors in Air Transport Design" page 535.
[2] Annex 6, par. 6.3.
[3] Annex 6, par. 4.2.7.5.
[4] Annex 6, par. 4.2.8.

a position to do so. In the case of forced landings at sea, the captains of other aircraft and surface craft are chiefly able to give such help.

THE BRUSSELS CONVENTION

In the Brussels Convention (1938) on assistance and salvage of and by aircraft at sea, an attempt has been made to lay down rules on the subject. This Convention is partly a matter of public law, because it contains a statutory obligation to render assistance, and partly private law because it gives rules concerning the remuneration for the services rendered. The Convention was signed by 16 countries, including the Netherlands, but there have not yet been any ratifications. [1]

According to Art. 2, both the commander of an aircraft and the master of a ship are bound to render assistance to anybody who is in danger of being lost at sea.

This obligation is greatly weakened, however, by the provision that "assistance" means any help which can be given, e.g. even a radio message is regarded as a form of assistance. Furthermore, the obligation does not apply if the (air)craft rendering assistance would be endangered, if the (air)craft is not under way or ready to depart, if it is not reasonably possible to render assistance or if assistance is already being rendered by others under similar or better conditions. [2]

It will be noticed that the Convention does not specify the period during which it is obligatory to render help, or in other words, the period during which it must be assumed that there may still be survivors. The length of time that one can keep alive on a rubber raft, for example, varies considerably, depending on the circumstances. [3]

With regard to remuneration, it has been stipulated that the indemnity shall not exceed 50,000 gold francs for each person saved (subject to a maximum total amount of 500,000 gold francs) or 50,000 gold francs if nobody is saved (Art. 3). In the case of salvage of an aircraft or of the objects on board, the remuneration

---

[1] Verschoor, "Het Verdrag van Brussel van 1938" page 9.
[2] See Verschoor, op. cit. page 28.
[3] For the factors playing a role herein, see McFarland, op. cit. page 540.

depends partly on the result ("no cure no pay"), the efforts and merits of the rescuers, and the value of the objects saved. The remuneration can never exceed the value of the goods salvaged (Art. 4). The remuneration will be divided between the operator and the crew in accordance with the national laws (Art. 6).

In judging the regulations with regard to the indemnities, one must take into account the possible ways in which aircraft can render practical assistance. When the Convention was drawn up the following methods were envisaged: [1]

1. Landing in the vicinity of the ship or aircraft in distress; this is only possible for flying boats, which are gradually becoming scarcer, and then only when the sea is comparatively smooth.
2. Dropping food, rescue equipment, etc.; as a rule this will only be possible for specially equipped rescue planes. The U.S. Air Force, for example, has aircraft which can drop a motor launch (furnished with provisions, first aid kit, radio, etc.) by parachute.
3. Tracing the scene of the accident and reporting the position; this will usually be the only way in which transport aircraft, for example, can assist.

Since the drafting of the Convention, the following manner of rendering assistance has also become important:

4. The special characteristics of the helicopter enable this type of aircraft to hover over the scene of the accident. Bij means of an electrically-operated tackle, people on the verge of drowning can be hoisted out of the sea. A number of spectacular rescues have already been performed in this manner [2].

As remarked above, the only way in which transport aircraft will normally be able to render assistance is by determining and if possible reporting the position. Any indemnity granted for this service relates solely to the costs incurred and any loss suffered during the rescue operations; as a general rule the whole of the indemnity will therefore be payable to the operator. As far as the commander of a transport aircraft is concerned, the rules relating to compensation and remuneration in the event of a rescue are therefore of little direct interest.

---

[1] Cf. Verschoor, op. cit. page 18.

[2] Cf. I. I. Sikorsky, "Helicopters in War and Peace", Journal of the Helicopter Association of Great Britain, Part V, No. 2, page 287; during the flood disaster in the Netherlands in February 1953, helicopters played a major part in rescue operations.

ANNEX 12 TO THE CHICAGO CONVENTION

ICAO has also drawn up regulations concerning search and rescue. These regulations, which are to be found in Annex 12 to the Chicago Convention, are based on different principles and cover a different field from the Brussels Convention.

The Annex is primarily intended to give effect to Art. 25 of the Chicago Convention, whereby the contracting States are obliged to take steps to render assistance to aircraft in distress on their territory, to co-operate in measures of assistance provided by the owners of the missing aircraft or the State in which the missing aircraft was registered, and, in general, to collaborate "in coordinated measures which may be recommended from time to time pursuant to this Convention".

Annex 12 therefore consists mainly of rules for the establishment of rescue services, the coordination of such services and the procedures to be followed. In addition, however, the Annex contains instructions directly addressed to the aircraft commander, viz. "Procedures for pilots-in-command observing an accident" (Par. 5.5.) and "Procedures for a pilot-in-command intercepting a distress call and/or message" (Par. 5.6).

DRAWBACKS AND ADVANTAGES

For purposes of comparison with the Brussels Convention, we shall now sum up the points of difference which may be regarded as drawbacks of Annex 12.

1. The first drawback is that the obligation to render assistance only applies to the commander of an aircraft and not to the master of a ship. This limitation is a consequence of the restricted scope of the Chicago Convention itself. To lay an obligation on masters of ships by means of an Annex to this Convention would be an inadmissible extension of its sphere of operation.

2. The arrangements do not apply for military aircraft but only for civil aircraft. The reason for this is similar to that explained above.

3. The obligation only applies if the aircraft commander "observes that either another aircraft or a surface craft is in distress." [1]

---

[1] Annex 12, par. 5.5.1.

This gives rise to two different comments.

a. Contrary to what is stated in the Brussels Convention, the aircraft commander is therefore not obliged to take off or to deviate from his course if he merely learns that an accident has taken place and does not observe this accident himself. On receiving a distress signal he can proceed to the scene of the accident "at his discretion" [1], i.e. there is no obligation to do so. In our opinion this is a very serious restriction, particularly because present-day transport aircraft usually fly at a very high altitude and often above the clouds, so it will be a remarkable coincidence if one observes an accident without altering course and keeping a special look-out.

b. Furthermore, the obligation only applies on observing an aircraft or surface craft in distress. We consider this wording rather unfortunate, for the aircraft commander is therefore not obliged to render assistance if he observes survivors from an aircraft or surface craft which has already sunk. The wording of the Brussels Convention (" ... assistance à toute personne se trouvant en mer, en danger de se perdre ...") is preferable in this connection.

4. The Annex does not lay down any rules for the remuneration. The absence of a ruling such as given in the Brussels Convention, which is mainly derived from sea law (where it may be of great importance), is in our opinion only a minor fault. [2]

To offset the above drawbacks we have the following advantages of Annex 12.

1. The obligation to render assistance also applies over land. This is very important because large areas of the globe are extremely inhospitable and sometimes help can only be provided by air. Needless to say, the risk of accidents over land is quite as great as the risk at sea. One need merely think of collisions with obstacles such as mountains, etc. CITEJA drew up a convention for assistance on land, but it never got beyond the stage of a preliminary draft. [3]

---

[1] Annex 12, par. 5.6.

[2] Otherwise, Riese, „Luftrecht" page 495; see also Goedhuis, "Handboek voor het luchtrecht" page 353.

[3] To be found in Verschoor, op. cit. page 93.

2. As already remarked, the chief way in which an aircraft can first render assistance is by making a reconnaissance and reporting the position of the scene of the accident. Annex 12 gives full details of the procedure to be followed by the aircraft commander: he must remain in sight of the distressed (air)-craft as long as possible; if his exact position is not known, he must determine it; he must supply the rescue coordination centre with information about the type of (air)craft in distress, its position, the time of observation, and the number and condition of the persons observed; lastly, he must comply with the instructions of the coordination centre. [1]

It is interesting to note that if no communication can be established with the rescue coordination centre, the commander of the first aircraft arriving on the scene is to take charge of the activities of all other aircraft that may arrive. [2] In this particular case the commander's authority therefore extends to other aircraft besides his own. As the commander of the first aircraft really takes the place of the coordination centre under such circumstances, he is entitled to give the same sort of instructions as might emanate from the rescue coordination centre.

The other aircraft are not obliged to comply with these orders, however, if they are not in the position to do so or if they consider the instructions unreasonable or unnecessary.

A final point of importance is that when an aircraft commander intercepts a distress signal, he must endeavour to plot the position of the (air)craft in distress or at least take a bearing on it. [3]

3. The help from ships and aircraft which happen to be passing can only be fortuitous and fragmentary. To ensure the success of a rescue attempt, systematic and coordinated action is generally necessary. Annex 12 specifies the way in which the measures of assistance must be coordinated when different countries are involved.

Without entering into a detailed discusion of the directives

---

[1] Annex 12, par. 5.5.1.
[2] Annex 12, par. 5.5.1.1.
[3] Annex 12, par. 5.6.

in question, it may be remarked that they relate to an organization which is of vital significance for air safety.

4. Annex 12 is already effective and that is a factor of practical importance. Probably it has not yet been introduced into all the national legislations but one may expect that this will take place in the near future; in the meantime the rules laid down in Annex 12 are already generally applied.

In the foregoing we have attempted to weigh the merits of the Brussels Convention and Annex 12 against each other. We consider that there is not much chance of ratification of the Brussels Convention in its present form, but it is desirable that the most useful features of the two documents should be combined in order to arrive at universal agreement on this chapter of air law which is so important for air safety.

Chapter VII

SANCTIONS

THE DUTCH CRIMINAL COURT

Infringement of statutory regulations may make the aircraft commander liable to prosecution. Naturally the general principles of criminal proceedings hold good here, but some difficulties are encountered in applying them.

To begin with, it seems that a criminal offence committed on board a Dutch aircraft *outside* Dutch territory is not indictable in the Netherlands (under the terms of Art. 2 of the Dutch Criminal Code). One might of course assume that Art. 3 of the Dutch Criminal Code, which only speaks of surface craft, is applicable to aircraft by analogy. [1] But this does not obviate the difficulty that the sphere of operation of the Aviation Act (and hence the orders issued under that Act) is expressly limited by defining air navigation as "the use of aircraft on and above the surface of the earth within the territory of the Netherlands." [2]

We understand that the draft of the new Aviation Act will omit the restrictive definition quoted above and — quite rightly, in our opinion — contain a proposal that Art. 3 of the Dutch Criminal Code should be amended to cover aircraft as well.

At present the scope of the Aviation Act is particularly limited as far as airline pilots are concerned. [3] This is obvious when one considers that Dutch airline pilots are engaged in air navigation outside the Netherlands as a rule rather than an exception.

In the course of a round trip from Amsterdam to Sydney (over 100 flying hours) the aircraft commander only spends one hour over Dutch territory.

---

[1] This point of view is defended by Pompe, ,,Handboek van het Nederlandse Strafrecht'' page 477.

[2] Article 1, sub. I.

[3] cf. Spanjaard, ,,Vliegwereld'', 6th April 1950, page 230.

As already pointed out, [1] the provisions of Dutch air legislation, especially those which directly concern the aircraft commander, are to a large extent out-of-date. This will inevitably lead to regular, if not systematic, breaches of the law by flying personnel who realize that the regulations in question are obsolete. To the best of our knowledge, legal proceedings are seldom instituted in such cases.

It must also be pointed out that it is by no means easy to detect offences in air navigation. This is primarily because one is dealing with a movable vehicle which, unlike a motor vehicle, cannot be observed at close quarters or stopped while in use. In many cases it will therefore be impossible to obtain proof of transgressions during the flight, unless they are revealed by the exchange of messages or subsequently discovered from the log books. Furthermore, it will often be necessary to have technical knowledge of aviation, which one cannot expect every police official to possess, [2] in order to judge whether an offence has actually been committed. One need only think of the difficulty of correctly estimating the altitude of an aircraft and deciding whether public safety is endangered by a certain manoeuvre.

Dutch air legislation contains a detailed set of rules of conduct for the aircraft commander. Different sanctions are imposed for violations of these rules, the maximum penalty being imprisonment for a period of one year or a fine not exceeding 3,000 guilders, together with suspension of the offender's licence for not more than six months. [3] For the reasons explained above, however, prosecutions and convictions of flying personnel are comparatively rare, in fact they are extremely rare in the case of airline pilots. Although the high selection standards, stricter discipline and greater responsibility obviously play an important part as far as the airline pilots are concerned, this does not mean that the latter are seldom guilty of infringing regulations.

In the exercise of his profession, the aircraft commander can be called to account not only for violations of air law but also for offences against common law.

This may occur, for example, when proceedings are instituted

---

[1] See page 16.

[2] Excepting personnel of the Department of Civil Aviation, who have been designated as investigating officials.

[3] Aviation Act, Article 44, par. 1. and article 47, par. 1.

on account of death or bodily injury through culpable fault, under the terms of Arts. 307–309 of the Dutch Criminal Code.

Many such cases with varying results have been dealt with in American courts, [1] but as cases of this nature have been extremely rare in the Netherlands we shall merely draw attention to the possibility of such proceedings without devoting further attention to the matter.

THE FOREIGN CRIMINAL COURT

As previously remarked [2], opinions differ in regard to the jurisdiction applying on board an aircraft. It is therefore impossible to say definitely what sanctions are applicable to the commander of a Dutch aircraft who commits an offence outside Dutch territory.

If an aircraft commander is guilty of a criminal act committed on or above the territory of another State, as a general rule he will be liable to criminal proceedings in that State. But the difficulty is that the time at which the offence took place cannot always be established with certainty — especially if the action in question occupied a certain space of time — and it is clear that the court having jurisdiction will depend on the position of the aircraft as indicated by the time of the occurrence.

Cooper has therefore proposed that the State of Registry and the State where the aircraft lands, as well as all the countries flown over, should be declared competent.

The person concerned will then be handed over — with due observance of the existing rules on the subject — to the country which first requests extradition under that regulation. [3] This appears to be a practical solution which may remove a lot of legal uncertainty. [4]

The above difficulty does not arise when air navigation regulations are violated, because in such cases the time of the occurrence and the position of the aircraft can usually be accurately determined.

---

[1] cf. Dijkstra, "Business Law of Aviation" page 491; Fixel, "The Law of Aviation", page 228; Manion, "Law of the Air" page 241.

[2] See page 32.

[3] I.L.A. Air Law Committee, Lucerne Conference 1952, page 2.

[4] The objections raised by Meyer, as indicated on page 26 of the cited report, do not appear to be convincing.

In this connection there is generally a dual jurisdiction, viz. that of the country where the aircraft is registered and that of the country where the incident takes place. [1]

An exceptional situation arises in flights over open sea. An American author [2] has posed the question: "Where ... could a common law crime, even murder, go unpunished in any court of law?" the answer being: "On a U.S. commercial airliner flying over the high seas". This is because an American court recently declared itself incompetent to give judgment in such a case. During a flight from Puerto Rico to New York two passengers had come to blows, and when the aircraft commander tried to separate them he was bitten in the shoulder and sustained injuries which caused loss of blood, while the stewardess also received some blows. As a result of the above decision the wrongdoers escaped punishment. There is some doubt as to whether the court's decision was correct, [3] but nevertheless it has been considered necessary to draft legislation which would change the existing law in order to meet this deficiency. As mentioned in an earlier chapter, [4] persons on board Dutch aircraft are still in a similarly "lawless" situation, so that it seems desirable to rectify matters by amending the law as soon as possible.

Although we do not share the view that a collison between two aircraft is "der im Deliktsrecht der Luftfahrt bedeutsamste und haüfigste Fall",a brief reference to this subject must not be omitted.

The general opinion is that the position here is closely analogous to that prevailing in maritime law [5].

After the French ship "Lotus" had collided with a Turkish vessel, on entering a port in Turkey the chief officer of the "Lotus" was arrested and sentenced to imprisonment by a Turkish court after being found guilty of manslaughter. On that occasion the Permanent Court of International Justice ruled that international

---

[1] This principle for example appears in article 12 of the Chicago Convention; this, however, will only apply to Netherlands aircraft when the Aviation Act is altered in the manner mentioned above (page 76).

[2] Hilbert, "Jurisdiction in High Seas Criminal Cases", JAL 1951, page 427 and 1952, page 25.

[3] Hilbert, loc. cit.

[4] See page 33.

[5] Frese, "Fragen des internationalen Privatrechts der Luftfahrt unter besonderer Beruecksichtigung einer Anwendungsmöglichkeit des Flaggenrechts", page 38.

law did not exclude the criminal jurisdiction of other countries in such cases. [1]

Although this verdict related to shipping, in our opinion there is no reason why a similar standpoint should not be adopted with regard to aviation. As far as the applicable law is concerned, this may be either the law in force at the scene of the collision or the national law of the aircraft.

According to Frese [2], who favours the latter solution, it will be necessary to fall back on the "law of the flag" for lack of any other starting-point in case of collisions over open sea or stateless territory, though complications may be anticipated if the two aircraft are of different nationalities and are subject to different systems of law. In a report drawn up for the Legal Committee of ICAO, however, the rapporteur Iuul considers that as a general rule either the law of the place where the accident took place, or the "lex fori" applies [3].

The scope of the present study does not permit us to go any further into this legal problem, which has by no means been solved. For our purpose it may suffice to remark that under existing air law in many cases the aircraft commander is in a state of uncertainty as to whether, and if so where and under what law, he may be tried for a punishable offence committed by him.

THE AIR ACCIDENT BOARD

Besides the criminal court, in the Netherlands there is another body which keeps a watchful eye on the activities of flying personnel. This body, the Air Accident Board, was set up pursuant to the Air Accident Act of September 10, 1936 (S. 522). Its main task is to investigate aircraft accidents, but it can also take certain measures against members of the crew, irrespective of whether an accident has occurred or not.

The Board consists of a chairman, who must be a laywer, and four members, preferably aviation experts. At least ten associate members are also appointed; they must be selected in such a way

---

[1] Cf. Ripert, ,,Précis de Droit Maritime" page 53.
[2] Loc. cit. page 38 onwards.
[3] Minutes and Documents Legal Committee, 7th Session Doc. 7157–LC 130, page 305.

as to include persons who hold or have held posts as an airport manager, aircraft commander, radio operator, aircraft constructor, electrical engineer, mechanical engineer, meteorologist or physician. If the chairman of the Board considers that the presence of one or more of the associate members is desirable on account of the nature of the case to be dealt with, he invites them to serve on the Board (Air Accident Act, Art. 1). In actual practice there is nearly always at least one member of the Board who is or has been a pilot (aircraft commander).

Naturally this may help to ensure that the special circumstances under which an aircraft commander has to perform his duties will be taken into account in the findings of the Board.

The Board is primarily concerned with accidents to civil aircraft, though it must be remarked that the term "accident" is given a very wide interpretation.

Accidents within the meaning of the Air Accident Act are occurrences resulting in serious bodily injury to one of the persons on board or any other person, in addition to occurrences in which the aircraft or any other property is seriously damaged. As far as Dutch aircraft are concerned, the term also includes any occurrence in which an aircraft is involved

1. if the safety of the persons on board is seriously endangered as a result of that occurrence;
2. if the nature of the occurrence is such that it will probably
   (a) be possible to draw lessons from an investigation or
   (b) found desirable to lay down regulations which may serve to prevent air accidents (Art. 5).

This wide definition allows the Board to take cognizance of occurrences of such a nature that a public inquiry may be of importance for air safety, even when no bodily injury or material damage has been caused.

The jurisdiction of the Board, in contrast to that of the Dutch criminal court, extends to occurrences involving Dutch aircraft outside the territory of the Netherlands.

The inquiry consists of a preliminary investigation by one or more persons designated for that purpose by the Minister of Transport and, if necessary, a further investigation by the Air Accident Board (Art. 6). The Director-General of the Netherlands Department of Civil Aviation has been designated by the Minister of Transport as the preliminary investigator within the meaning of the Air Accident Act. In the preliminary investigation the Director-General may be represented or assisted by some officials — also appointed for this purpose — of the Department of Civil Aviation.

The powers of the Air Accident Board in regard to crew members are firstly, to judge their competency to perform duties on board an aircraft (Arts. 22–29) and, secondly to take disciplinary measures (Arts. 37–40).

DECLARATION OF INCOMPETENCY

Strictly speaking, judgment as to the competency of crew members by the Air Accident Board cannot be regarded as a sanction. Since it is sometimes difficult to distinguish between a measure taken under this head and a disciplinary measure, apart from the fact that such a measure may have extremely serious consequences for the person concerned, we shall briefly review the position.

An inquiry into the competency of a crew member can take place not only if an accident has occurred, but also — in special circumstances — if no accident has occurred.

In the first case, if the inquiry into the circumstances of an air accident reveals facts which make the preliminary investigator (i.e. the Director-General of the Department of Civil Aviation) wonder whether a member of the crew is really competent, his report to the Board will be accompanied by a recommendation that the crew member in question should be heard (Art. 22, Par. 1).

As commented above, the definition of the term "accident" in the Air Accident Act is very comprehensive. Notwithstanding this, provision is also made for an inquiry into the competency of a crew member even when an accident has *not* occurred; in such a case the Director-General of the Department of Civil Aviation can submit a proposal to that effect to the Air Accident Board (Art. 22, Par. 2).

Pending the inquiry, the Board may temporarily suspend the licence of the person concerned (Art. 22, Par. 8).

After investigation the Board may declare a crew member incompetent to serve in a certain capacity on board an aircraft. This can be done either at the instance of the Director-General of the Department of Civil Aviation or on the Board's own initiative, and the reasons for such a decision will also be stated. A certificate of competency issued to the person concerned then ceases to be valid on the date when the Board announces its findings (Art. 24).

It is important to note that the Board's judgment of the competency of a crew member need not necessarily be based on the same standards as those which the Director-General is bound to apply. According to the explanatory memorandum appended to the Air Accident Act, "even though the person concerned has passed the appropriate examinations, if the Director-General still considers that the said person is incompetent he can take the matter up with the Air Accident Board in conformity with Art. 22 (2). It should be remembered that in judging whether the person concerned is "incompetent" the Board does not have to adhere to the requirements laid down in the R.T.L. with regard to physical fitness and skill."

DISCIPLINARY MEASURES

The disciplinary measures which the Air Accident Board can take against crew members consist of a reprimand or suspension from performance of the duties specified in the certificate of competency for a stated period (not more than two years). The Board can impose a disciplinary punishment of this nature — without prejudice to civil and criminal proceedings — if the inquiry into an accident has convinced the Board that the accident was due to a fault on the part of a crew member (Art. 37). This means that in the latter case the jurisdiction of the Board does not go so far as in the "declaration of incompetency," since the Board can only impose a disciplinary punishment if there is question of an accident within the meaning of the Air Accident Act — which certainly allows a lot of scope — while the element of "fault" is also introduced.

Apart from the question of whether the term "fault", which suggests criminal proceedings, is appropriate in proceedings such as those of the Air Accident Board, the wording used may give rise to difficulties. After all, an air accident is often attributable to several different factors (e.g. bad weather, engine trouble and faulty instructions from the ground services together with "fault" of the aircraft commander), so that in many cases one can only say an accident was *partly* due to the fault of a crew member. Since Art. 37 states that possible disciplinary measures taken by the Board are independent of civil and criminal proceedings, the legislator obviously did not intend to give the Air Accident Board any powers of criminal jurisdiction. A double criminal prosecution for one and the same act would naturally be contrary to legal principles. Moreover, it is expressly stipulated in Art. 36 — perhaps superfluously — that if a criminal offence has been committed the person in charge of the investigation must immediately notify the Public Prosecutor.

As far as the aircraft commander is concerned, the disciplinary power of the Board also extends to cases in which *no* accident has taken place. Under the terms of Art. 37(2), the Board can take disciplinary action against an aircraft commander (or his deputy) who has been guilty of misconduct in any way whatsoever towards his company, the shippers or the persons on board the aircraft, or who has in any way failed to fulfil his statutory obligations.

Here we are really referring to offences which will only be prosecuted on a complaint being made. A complaint may be investigated if it is lodged by the Director-General of the Department of Civil Aviation, by the company employing the aircraft commander (or his deputy), or by one or more of the underwriters, shippers or persons on board the aircraft. The complaint cannot be entertained unless it is submitted to the Director-General of the Department of Civil Aviation or the Air Accident Board within three months after the date on which the complainant learned of the act in question.

The rules are analogous to the corresponding provisions which apply to the master of a ship, and they are based on the exceptional and responsible position held by the aircraft commander as well as the ship's master. [1]

SAFEGUARDS FOR THE PERSONS CONCERNED

In view of the far-reaching powers of the Board with respect to crew members, it is only right that the procedure should include a number of safeguards for the persons concerned. Some of these provisions are mentioned below:

---

[1] Cf. Cleveringa, ,,Het Nieuwe Zeerecht'' page 240.

If a Committee of the Air Accident Board decides that a crew member must be heard, a copy of this decision is served upon the person in question (Art. 22, Par. 3).

If it is first decided at a session of the Board that the competency of a crew member shall be investigated, the person concerned is duly notified; the further proceedings may then be postponed at his request (Art. 22, Pars. 4 and 5). If the person concerned is not present at the session, the hearing of the case is suspended (Art. 22, Par. 6).

The person concerned can arrange to be assisted by an adviser or represented by a proxy. The person concerned, and his adviser or proxy, are entitled to examine the documents relating to the preliminary investigation at the secretariat of the Board prior to the hearing of the case (Art. 22, Par. 7).

The person concerned can ask for another hearing of the earlier witnesses or other persons designated by him (Art. 23). If a case has been heard in the absence of a crew member who is declared incompetent (or his proxy) he can lodge written notice of appeal within a certain period after being informed of the Board's findings (Art. 25, Par. 2). This appeal will then be considered by the Board and it is possible that the case may be reopened (Art. 26).

As may be seen from the foregoing, there are many provisions purporting to safeguard the interests of the persons concerned. This is necessary, in view of the importance of the interests which may be at stake — especially for those who have made flying their profession.

It is regrettable, however, that in spite of this the Air Accident Act is seriously defective in regard to the legal remedies open to the person on whom the penalties are imposed.

Admittedly Art. 28 provides for a fresh inquiry by the Board "when new facts come to light which were not yet known when the case was heard and which might have affected the decision". Under the terms of Art. 29 the Board can also restore all or part of the competency which the person concerned was deprived of, if it may be assumed that he is once again competent to perform his duties on board an aircraft. In addition, after consulting the Air Accident Board the Crown can either restore all or part of the competency which the person concerned was deprived of or else reduce the period of incompetency (Art. 40). Nevertheless, the person concerned cannot appeal to a higher Court.

The Air Accident Act is copied practically word for word from the Dutch Merchant Shipping Act, where a similar procedure is followed in shipping inquiries. In the latter case too the person concerned has no opportunity for appeal. This shortcoming, like other faults in the course of procedure, has already been pointed out in the relevant literature. [1]

We are in favour, as a matter of principle, of introducing into the Air Accident Act the possibility of appeal to a higher Court. On the other hand, it should be mentioned that in actual practice the composition of the Board helps to guarantee to a large extent that justice will be administered correctly in most cases, while it has also been found that the Board makes very limited use of its far-reaching powers in regard to flying personnel.

During the years 1945–51 the Air Accident Board only once declared a crew member to be incompetent under the terms of Art. 24 of the Air Accident Act. The man in question was a private pilot and the Board's findings were that, "on account of his qualities of character he must be deemed incompetent to exercise his function as an aircraft pilot."

In the same period the Board took disciplinary measures under Art. 37 in sixteen cases. In eight of these cases the offenders were reprimanded and in the other eight cases the licences of the persons in question were suspended for periods varying from a fortnight to the maximum of 2 years.

Through such corrective measures the Air Accident Board makes a valuable contribution towards air safety.

One very important point is that the Air Accident Board does not have to confine itself to the question of whether statutory regulations have been violated or not, when passing judgment on crew members. For example, the Board can impose sanctions if it considers that an aircraft commander has not observed the care which is essential in air navigation, if a generally accepted rule or usage in air navigation has been contravened, or if ICAO regulations have not been complied with (even though they have not yet been given force of law). The latter is particularly impor-

---

[1] See Wolfsbergen, "Processuele curiosa bij de Raad voor de Scheepvaart", NJB 1938, page 82.

tant in view of the fact that, as already remarked, many of the ICAO regulations have not yet been introduced into the national legislation of the Netherlands. [1]

In this respect the Air Accident Board therefore performs a task which may to a large degree supplement that of the Criminal Court.

---

[1] See page 17.

CHAPTER VIII

LIABILITY

A. THE WARSAW CONVENTION

The liability of the international air carrier is governed by the "Convention for the unification of certain rules relating to international carriage by air,", which was signed at Warsaw on October 12, 1929. This Convention was ratified by the Netherlands on April 6, 1933 (S. 149) and came into operation on September 29, 1933.

Before discussing whether, and if so to what extent, the rules of liability laid down in the Convention are applicable to the aircraft commander, we shall outline the present arrangements. [1]

The Convention applies to international carriage of persons, baggage or goods performed by aircraft for reward. In the Conditions of Carriage adopted by the members of IATA, [2] however, the rules of the Convention (with a few exceptions) are likewise declared applicable to international carriage not covered by the Convention, and also to non-international carriage. The scope of the Convention was similarly extended by the Dutch Act of September 10, 1936, which decreed that the rules set out in the Convention should apply to air transportation inside the Netherlands as well as between the Netherlands and a country which has not ratified the Convention. Otherwise the text of this statute is practically identical to that of the Convention. [3]

Chapter II deals with the documents of carriage and lays down rules concerning passenger tickets, baggage checks and air consignment notes, while Chapter III relates to the liability of the carrier. The Convention establishes the principle of liability,

---

[1] The Legal Committee of ICAO keeps itself intensely occupied with pro posals for revision of the Convention.

[2] IATA Conditions of Carriage A art. 1. and B. art. 1.

[3] For an exception see page 94.

fixes the limits of liability, and makes provision for cases in which the liability is either non-existent or unlimited, as may be seen from the following.

The carrier is liable for damages in the event of a passenger being killed or wounded or sustaining any other bodily injury as the result of an accident, and the same applies in the event of destruction, loss or damage of registered baggage or goods during a specified period (Arts. 17 and 18).

The carrier's liability is limited to a sum of 125,000 gold francs for each passenger, 250 francs per kilogramme of goods and registered baggage, and 5,000 francs per passenger for baggage kept in the passenger's own custody (Art. 22).

The carrier is *not* liable if he can prove that he and his servants took all necessary measures to avoid the damage, or that it was impossible for him or them to take such measures.

With regard to the carriage of goods and baggage, the carrier is not liable if he can prove that the damage was due to an error in the pilotage, in the handling of the aircraft or in the navigation, and that in every other respect he and his servants took all necessary measures to avoid the damage (Art. 20).

The carrier is not entitled to avail himself of the provisions of the Convention which exclude or limit his liability — in other words the carrier's liability is *unlimited* — if the damage is caused by his wilful misconduct or any fault on his part which, according to the law of the court seised of the case, is considered to be equivalent to wilful misconduct. The carrier's liability is also unlimited if the damage is caused under the same circumstances by one of his servants acting within the scope of his employment (Art. 25).

We shall not go into the various other provisions governing the extent of the liability, such as those in which the carrier is threatened with the sanction of unlimited liability if he fails to issue a passenger ticket, baggage check or air consignment note, or if these documents are not fully completed (Art. 3, Par. 2; Art. 4, Par. 4; Art. 9). Another case which we shall leave out of consideration is the one where an injured person was partly responsible for an accident, with the result that the carrier is therefore wholly or partly exonerated from liability (Art. 21).

In some countries an injured party has the option of instituting an action for damages *ex contractu* or *ex delicto* [1]. To obviate the consequent legal insecurity for the carrier, Art. 24 provides that in the cases covered by Arts. 17, 18 and 19 an action can only be brought subject to the conditions and within the limits laid down by the Convention.

Applicability to the aircraft commander

The question to be answered now is whether the Warsaw Convention regulates the liability of the carrier's servants and of the aircraft commander in particular. In other words, can an aircraft commander avail himself of the provisions of the Convention which would regulate, limit or exclude his liability, or can he be held liable under the national law, e.g. in an action for tort, without any limitation of his liability?

If the aircraft commander is also the carrier, as may well occur in the case of one-man companies, in our opinion there is no doubt that through pleading that he is the carrier, he can only be held liable subject to the conditions and limits set out in the Convention.

The position is different when the aircraft commander and the carrier are not the same person; such cases occur more frequently and are therefore of greater practical importance. Opinion is by no means unanimous in regard to the latter cases.

Lemoine [2] considers that the provisions of the Convention are indeed applicable to the carrier's servants. In an article on the desirability of revising the Warsaw Convention, however, Beaumont [3] mentions the view that employees — including aircraft commanders — can be sued separately and that they are not protected by the limitations of liability provided by this Convention. Maschino [4], Riese [5], Bucher [6] and Koffka [7] concur with this point of view.

---

[1] Cf. Goedhuis, "National Airlegislations and the Warsaw Convention", page 267.
[2] Lemoine, "Traité de Droit Aérien" No. 822, 840 and 841.
[3] Beaumont, "Need for revision and amplification of the Warsaw Convention", JAL 1949, page 395.
[4] Maschino, "La Condition Juridique du Personnel Navigant de l'Aéronautique", page 125.
[5] Riese, "Luftrecht", page 441.
[6] Bucher, "Le Statut juridique du personnel navigant de l'Aéronautique", page 36.
[7] Koffka — Bodenstein — Koffka, "Lufverkehrsgesetz und Warschauer Abkommen", page 269.

In the writer's opinion, the latter view is correct. Both the previous history and the preamble of the Warsaw Convention show that the intention was to regularize the liability of the *carrier*; there is nothing to indicate that the rules were also intended to apply to the carrier's *servants*.

This is confirmed by the text, since Arts. 17, 18 and 19 (which establish the principle of the liability) and Art. 22 (which lays down the limits of this liability) only speak of "le transporteur" and not of "les préposés", quite apart from the question of whether the aircraft commander should be regarded as belonging to the latter category [1]. This is all the more striking because the carrier's servants are mentioned in other Articles, e.g. Arts. 20 and 25.

Art. 25 is worth noting in this connection since the first paragraph deals with cases where the *carrier* has been guilty of wilful misconduct or equivalent default, while the second paragraph gives an identical ruling for cases where the carrier's *servants* have been similarly guilty within the scope of their employment. This contradicts the theory that the servants are covered by the term "the carrier" elsewhere in the Convention, and leads us to infer that the rules of liability set out in the Convention do not apply to the aircraft commander. If the latter view is accepted, in certain circumstances a carrier may be wholly or partly exonerated from liability for damage attributable to an error in navigation, under the terms of Art. 20(2) or a similar indemnity clause in the national legislation, whereas in such a case the aircraft commander may be held liable without any limitation. Verdicts of this kind actually exist in jurisprudence. [2]

One may agree with Lemoine [3] that the above situation does not satisfy one's sense of justice. It is true that in the IATA Conditions of Carriage [4] all liability of employees is excluded, but doubt has been expressed as to whether a hold-harmless clause of this nature will be deemed admissible by the court seised of the case. [5]

The object aimed at in drawing up the Convention was to regu-

---

[1] Regarding this question see Goedhuis, "National Airlegislations" page 224; Bucher, op. cit. page 36; van Houtte, op. cit. page 88.

[2] Mathon, Mourier et Nigaij v. Brutschy et Société Caudron, RGDA 1937, page 148 and 1938, page 91; cf. Riese, op. cit. page 441; Lemoine, op. cit. No. 840.

[3] ibidem. No. 841.

[4] IATA Conditions of Carriage, art. 18, par. 5.

[5] Riese, op. cit. page 441.

late and limit to some extent the carrier's liability, on account of
the uncertainties of aviation. This was indeed achieved. But the
same consideration applies all the more for the employees of
the carrier, since they are even more vulnerable from a financial
viewpoint. In the event of an aircraft commander being killed in
an accident, it will often be very difficult for his relatives to prove
that the accident was not due to his fault; an unfavourable verdict can have very serious financial consequences in view of the
huge sums which may be involved in claims for compensation.
IFALPA, the International Federation of Airline Pilots Associations, has already expressed deep concern about the lack of statutory rules of liability for aircraft commanders, especially rules
adapted to the special circumstances prevailing in air transport [1].
In our opinion this point must receive full attention when the
Warsaw Convention is revised.

A subcommittee on the revision of the Warsaw Convention has
recently drafted an amended Convention which will be discussed
at the 9th Session of the Legal Committee of ICAO in 1953. Art.
13b of the proposed new text goes far to meet our objections
against the present rules as far as the liability of the aircraft
commander is concerned. The clause in question reads as follows:
"... in so far as servants and agents of the carrier are personally
liable for damage arising under any of the circumstances provided
for by this Convention, they shall be entitled to the same limits of
liability as those applicable to the carrier". [2]

From the explanatory notes, however, it appears that there is
a cleavage of opinion as to the desirability of such a clause. In
our opinion inclusion of a provision of this tenor in the Convention is higly desirable. A clause to this effect has recently been incorporated in the Rome Convention. [3]

Liability in the event of delay

The regulations concerning liability in the event of delay also
call for some comment. These regulations can be of importance
to the aircraft commander because in actual practice he may of-

[1] ICAO Doc. A4 — WP/154 and The Aeroplane, 2nd June 1950, page 645;
cf. also Honig, "De Positie van de Gezagvoerder van een Luchtvaartuig", NJB 1951,
page 317.
[2] L. C. Working Draft 391, page 15.
[3] See page 96.

ten have to take decisions which will cause delay. Examples of this are diversion to an airport other than the intended destination in view of unfavourable weather conditions, making an unscheduled landing on account of engine trouble, leaving some of the passengers or cargo behind in order to avoid overloading, etc.

According to Art. 19 the carrier is liable for damage occasioned by delay in the carriage by air of passengers, baggage and goods. The inclusion of this provision was at first strongly opposed by the airlines on the ground that the then existing state of technical development (in 1929) was such that a liability of this nature could not be accepted in actual practice. On the other hand, however, it was argued that travellers and shippers usually choose air transportation with the object of saving time. If this benefit were to be lost through delay, it did not seem right to exempt the carrier from responsibility for resultant damage. [1]

In judging the question of what constitutes delay, it is important to know what has been agreed upon between the parties to the contract of carriage. In the conditions of carriage adopted by the members of IATA it is expressly stated that the times shown in the timetables are not guaranteed. [2]

According to Art. 23 of the Convention, however, any provision tending to relieve the carrier of his liability or to fix a lower limit than that laid down in the Convention is null and void. This means that the carrier cannot reduce or preclude his liability under the terms of the contract of carriage. It may be assumed, however, that in deciding what is to be understood by delay the court will take into account the conditions of carriage and will ignore a single failure to adhere to the times shown in the timetable. Should these times be unduly exceeded, then the carrier may be held liable, though in that case he can still avail himself of the provisions of Art. 20, [3] whereby he is not liable if he can prove that both he and his servants took all necessary measures to avoid delay, or in other words, that his organization functioned properly. This may occur, for example, if delay is caused by unexpectedly bad weather conditions.

If a pilot forgets his charts or if the aircraft is not refuelled in

---

[1] Riese, "Luftrecht", page 448, 449.

[2] IATA Conditions of Carriage art. A. 19 and B. 20.

[3] Goedhuis, ,,National Airlegislations and the Warsaw Convention", pages 208,209; Riese, "Luftrecht", pages 450, 451.

time, with the result that there is an abnormal delay and consequent damage, then the carrier may very definitely be liable. In
actual practice, however, there have not been any cases where an
action for damages on this ground was successful. Thus the appre-
hension of the airlines that the provisions of Art. 19 might give
rise to a large number of lawsuits was unjustified [1].

A further point to be noted is that Art. 28 of the Dutch law of
September 10, 1936, which contains certain provisions relating
to air transportation, allows the carrier to contract out of liability
for delay, notwithstanding Art. 23 of the Convention [2].

The question of delay is also broached in the proposal for revision of the Warsaw Convention. In this draft the carrier is held
liable for damage due to delay, except in the case of delay
" . . . for the purpose of saving life, or for reasons of safety or on
account of meteorological conditions, or other reasonable deviation on technical grounds". [3] Amendment of the Convention in
this sense appears advisable, although it will probably be a difficult task for the court to determine what is meant by a "reasonable deviation on technical grounds." [4]

B. THE ROME CONVENTION

A "Convention for the unification of certain rules relating to
damage caused by aircraft to third parties on the surface" was
signed at Rome on May 29, 1933. Originally it was only ratified
by five countries, viz. Belgium, Brazil, Guatemala, Rumania and
Spain.

The outbreak of World War II, as well as disagreement about
the manner in which security would have to be provided, stood
in the way of universal acceptance. Revision of the Convention
has been under consideration by the Legal Committee of ICAO
for some years past, and a new and completely revised version of
the Convention was adopted at a first international conference
on private air law held in Rome during September and October,
1952; no ratifications had taken place at the time of writing.

---

[1] Riese, op. cit. page 451.
[2] See Goedhuis op. cit. page 215.
[3] L. C. Working Draft 391, page 18.
[4] For the term "deviation" in maritime law in accordance with the Treaty of Brussels (1923), see Cleveringa, ,,Het Nieuwe Zeerecht", page 411 and the literature given
therein.

In the following pages we shall briefly discuss the principles of the revised Convention, insofar as they are of importance to our subject.

The idea underlying the Convention is that persons who are not concerned with the performance of air navigation must be entitled to compensation for damage caused by an aircraft. This follows from Art. 1 of the Convention, which reads as follows: "Any person who suffers damage on the surface shall, upon proof only that the damage was caused by an aircraft in flight or by any person or thing falling therefrom, be entitled to compensation as provided by this Convention. Nevertheless there shall be no right to compensation if the damage is not a direct consequence of the incident giving rise thereto, or if the damage results from the mere fact of passage of the aircraft through the airspace in conformity with existing air traffic regulations". On the other hand, the operators are protected against excessive risks by a limitation of their liability. This limitation depends on the weight of the aircraft. If the weight of an aircraft is 50.000 kilos, for example, the maximum liability will be 10.500.000 gold francs.

The liability in respect of loss of life or personal injury shall not exceed 500.000 gold francs per person (Art. 11). The liability can be diminished or set aside if the injured party has caused or contributed to the damage (Art. 6).

There is no limitation of liability, however, if the injured party proves that the damage was caused by "a deliberate act or omission of the operator, his servants or agents, done with intent to cause damage; provided that in the case of such act or omission of such servant or agent, it is also proved that he was acting in the course of his employment and within the scope of his authority" (Art. 12).

Lastly, the liability for payment of compensation falls on the operator. The term "operator" is defined at some length, but the main point to note is that it covers cases where the operator is using the aircraft personally or when his servants or agents are using it in the course of their employment, whether or not within the scope of their authority (Art. 2).

Applicability to the aircraft commander

As with the Warsaw Convention, we shall now ascertain

whether, and if so to what extent, the liability of the aircraft commander is regulated by the Rome Convention.

Under the terms of the Convention, the "operator" of the aircraft is liable in the first instance but it is quite possible that the aircraft commander may at the same time be the "operator" as defined in the Convention, e.g. in the case of one-man companies, or if a private pilot hires an aircraft from an aero club [1].

Should damage be caused to third parties on the surface in such cases the Convention will apply to the aircraft commander. As a general rule, however, the aircraft commander and the operator will not be one and the same person.

Prior to the recent revision, the Convention only governed the liability of the operator, c.q. the owner; the original draft contained a clause providing for liability on the part of the person causing the damage, but this clause was subsequently deleted [2]. It may rightly be inferred from this that the intention of the authors of the Convention was merely to regulate the liability of the operator, so that the liability of the aircraft commander for a wrongful act remained undiminished [3].

Naturally objections have been raised against this, and the revised Convention now contains an article (Art. 9), reading as follows:

"Neither the operator, the owner, any person liable under Article 3 or Article 4, nor their respective servants or agents, shall be liable for damage on the surface caused by an aircraft in flight or any person or thing falling therefrom otherwise than as expressly provided in this Convention. This rule shall not apply to any such person who is guilty of a deliberate act or omission done with intent to cause damage".

We see from this article that, just as in the Warsaw Convention, the airline can only be held liable in accordance with the rules of the Convention, unless there is wilful misconduct involved.

Furthermore — and this addition is very important — the aircraft commander can never be sued by third parties, except in the event of wilful misconduct or unlawful use of the aircraft.

---

[1] Cf. Goedhuis, Handboek page 301.
[2] ibidem page 302.
[3] Goedhuis, Handboek, page 304; Oppikofer, "Internationale Luftprivatrechts-konferenz", ArchfLR 1933, page 225; Riese, "Luftrecht", page 349; contra: Coquoz, "Le Droit Privé International Aérien", pages 178, 208.

This is because the Convention states that the only parties who may be held liable are the operator and, under certain circumstances the owner or person who "wrongfully takes and makes use of an aircraft without the consent of the person entitled to use it". The above provision of the Convention serves to vindicate our point of view that the aircraft commander, who through the nature of his profession is exceptionally liable to make mistakes which can cause serious damage, ought to be safeguarded against the financial consequences of such errors. We entirely agree that this exclusion of liability should not apply if there is any question of wilful damage.

But the Convention also includes a provision (Art. 10), the effect of which is to make the above-mentioned safeguard null and void to a large extent; the article in question reads:

"Nothing in this Convention shall prejudice the question whether a person liable for damage in accordance with its provisions has a right of recourse against any other person."

This means that if damages are awarded against a carrier on account of an error in pilotage, the carrier can seek recourse against the aircraft commander.

Although the larger airlines will very seldom avail themselves of this right of recourse against the aircraft commander — insofar as they have not previously renounced their possible right of action in such cases — there is still a chance that the commander may find himself reduced to poverty through a single incorrect manoeuvre. We therefore think it advisable that the convention on the legal status of the aircraft commander should include a provision dealing exhaustively with this subject. [1]

Damage by persons on board

Prior to the recent revision, the Convention contained a clause (Art. 2, Par. 2b) defining damage within the meaning of the Convention as: "Le dommage causé par une personne quelconque se trouvant à bord de l'aéronef, sauf dans le cas d'un acte intentionnellement commis par une personne étrangère à l'équipage, en dehors de l'exploitation sans que l'exploitant ou ses préposés aient pu l'empêcher".

---

[1] See page 158.

7

This saving clause allowed the airline to plead inability to prevent damage caused by a passenger. Goedhuis [1] cites the instance of a passenger in an aircraft firing at a person on the ground. In such a case it is possible that the airline could have taken action to prevent the passenger from carrying a loaded firearm on board the aircraft — which is forbidden by law [2] — but under certain circumstances the operator might well allege that the aircraft commander was unable to take action against the passenger through lack of adequate powers. Following the revision, however, this clause has disappeared from the Convention; in other words, the carrier is liable for damage caused to third parties by persons on board the aircraft, although he can seek recourse against the person responsible for the damage. This seems to be an argument for granting the aircraft commander legal authority over the persons on board, so that he can deal with recalcitrant or dangerous passengers when the need arises. Otherwise it would be most unfair to hold the operator liable for damage to third parties in cases of this nature.

C. PILOT ERROR

A further study of what constitutes "pilot error" is important for two reasons: firstly, because accidents are often ascribed to pilot error, and secondly, because the meaning of this term affects the liability of both carrier and aircraft commander.

"Mistakes made by pilots and aircrew members are among the most frequent of accident causes" [3] according to a manual issued by ICAO.

The C.A.B. has announced that in non-aircarrier flying during 1948 90% of the total of 850 fatal accidents were attributable to pilot error [4].

In commercial air transportation this percentage is lower on account of better training, selection and discipline of the flying personnel. Nevertheless in 1949 pilot error was the primary cause in 55% of all notifiable air accidents in the U.K. [5]. McFarland

---

[1] Handboek, page 296.

[2] R.T.L. art. 196.

[3] Manual of Aircraft Accident Investigation, ICAO Doc. 6920 — AN/855, page 73.

[4] CAB report "The Human Equation in Aircraft Accidents" cited in Interavia Air Letter No. 1915, 14th March 1950, page 4.

[5] Group Captain J. Veal, "Some British Views on Flight Safety Measures", Third Annual Safety Seminar, Flight Safety Foundation, October 30th, 1950.

arrives at a figure of 39.8% for accidents due to pilot error on domestic air services in the United States, as compared with 31.2% for the military air transport services [1]. From statistics published by the C.A.A. it appears that 44.9% of the accidents in scheduled domestic air transportation and 49.4% of the accidents in international transportation were attributable to pilot error during the 1938–1946 period [2]. ICAO even reports a percentage of 62.3% for the 1949–1950 period and 79% for 1950–1951 [3].

Whatever one may think of these figures — bearing in mind the difficulty of establishing with absolute certainty the cause of an accident [4] — there is little doubt that aircraft accidents are frequently due to an action of the pilot.

This conclusion differs from the view mentioned by Goedhuis [5] and Van Houtte [6] in connection with the Warsaw Convention, namely that accidents are relatively seldom attributable to errors in pilotage. This difference of opinion can be explained by differences in the statistical methods or by the technical development of aviation since the Warsaw Convention was signed. It is true that technical failures are steadily decreasing, but on the other hand the greater speed and complexity of the aircraft impose heavier and heavier demands on the pilots [7]. As Ogburn [8] remarks: "Medical experts claim that there are physiological and psychological limits to the number of operations which one person can do, and suggest that the operations involved in piloting a plane must be simplified. Such a simplification would undoubtedly decrease the number of accidents now considered caused by pilot error."

Criticism of frequent use of term

The frequent use of the term "pilot error" to account for air accidents has given rise to criticism.

---

[1] McFarland, "Human Factors in Air Transport Design", page 602.
[2] CAA Statistical Handbook of Civil Aviation 1949, page 101.
[3] Aircraft Accident Digest No. 1, ICAO Circular 18-AN/15, page 9.
[4] cf. McFarland, op. cit. page 601.
[5] Goedhuis, "National Airlegislations and the Warsaw Convention", page 233.
[6] van Houtte, "La Responsabilité Civile dans les Transports Aériens Intérieurs et Internationaux", page 93.
[7] ICAO circular 18 — AN/15, page 11.
[8] Ogburn, "The Social Effects of Aviation", page 381.

In the first place it has been pointed out that it is often impossible to determine the cause of an accident with certainty, especially if there are neither survivors among the crew nor expert witnesses on the ground. There is a lot of truth in the above comment.

In this connection McFarland [1] remarks: "After each transport accident an attempt is always made to allocate the causes or place the responsibility on some definite factor such as the pilots, other personnel, the aircraft, the ground and navigational facilities, or the regulations governing the operations. Each of the interests represented naturally tries to establish its lack of responsibility for the crash ... The reconstruction of what happened during the crash by a post-accident analysis is subject to many errors and is often complicated by the efforts of these various groups to protect their special interests."

The cause will frequently have to be established by a process of elimination. For example, if an examination of the wreckage does not reveal any indication of technical defects, if the ground stations have functioned properly, if the aircraft was given correct information, if the aeronautical charts are found to be in order and if there are no other apparent causes, one may be inclined to blame the pilot when an aircraft has gone off its course [2]. However great this probability may be, one can never rule out the possibility that entirely unforeseen circumstances may have obliged the pilot to follow such a course. In view of this, one may adopt the same attitude as those who are unwilling to attribute an accident to pilot error unless it is absolutely certain, particularly in cases where the pilot is no longer able to speak for himself. Most reports on inquiries into aircraft accidents are therefore very cautiously worded as to the cause of the accident.

Another criticism is based on the idea that "pilot error" implies default, failure to do one's duty, or even negligence. In our opinion this idea rests on a misconception, a failure to distinguish between establishing the cause of an accident in a technical and statistical sense, and determining the responsibility in a legal sense.

In the words of one lawyer who is also an experienced pilot:

---

[1] op.cit. page 601.
[2] cf. McFarland, op.cit. page 567.

"Pilot error may, and often does, exist without legal responsibility of the pilot for the accident, because every act of the pilot
contributing to the accident although unavoidable or completely
justifiable, is called 'pilot error'. The accident analyst must
attribute the accident to aircraft structure, power plant, pilot,
other personnel, etc., and he is not concerned with whether the
pilot's action was justified or wrong in either a moral or legal
sense." [1]

In order to meet the above objection, the C.A.B. no longer
employs the expression "pilot error" to describe the cause of
individual accidents, but only to indicate the character of groups
of accidents for statistical purposes. [2]

It has also been suggested that for psychological reasons the
term "pilot error" should only be employed if there is any question of negligence or default on the part of the pilot. In other cases
the term "pilot action" should be used. [3]

Even if we take the term "pilot error" in the wide statistical
sense mentioned above, it will be found that opinion is divided as
to the meaning of this term. What standard is one to apply in
deciding whether a pilot's action was an "error"?

According to one line of thought, the standard of comparison
should be the action which a hypothetical "pilot of high skill and
prudence" might have been expected to take under similar circumstances [4]. If the result of such comparison proved to be
unfavourable, then there would be a question of pilot error.

On this subject the following comment has been made: "The
frequency with which accidents have been attributed to pilot
error may be due to the belief that the airman is able to do all the
things expected of him, i.e. that he is a 'perfect' pilot. Although
high standards of physical fitness, training and judgment are
required of transport crews, the concept of an average aircraft
pilot is a more reasonable assumption not only in the attribution
of blame for an accident, but also in the design of the plane" [5].

For this reason the C.A.B. has ruled that a pilot's actions and

[1] Flight Safety Foundation, Accident Prevention Bulletin 50–20, August 28th, 1950.
[2] idem 51–1, January 18th, 1951.
[3] idem 50–20, August 28th, 1950.
[4] Dr. A. D. Tuttle, Journal of Aviation Medicine, June 1939, cited in Accident
Prevention Bulletin 50–15.
[5] McFarland, op. cit. page 570.

decisions must be evaluated against "reasonable performance which normally could be expected from a pilot with equivalent certificate and experience" [1].

We prefer the latter criterion because it does not impose unreasonably high demands on the pilot.

Connection with liability

A clear definition of "pilot error" is extremely important for the carrier's liability.

The relevant provisions of the Warsaw Convention (Arts. 20 and 25) speak of "negligent pilotage or negligence in the handling of the aircraft, or in navigation" as well as "wilful misconduct or such default as, in accordance with the law of the Court seised of the case, is considered to be equivalent to wilful misconduct."

In the first case the carrier is not liable for damage in the transportation of goods and baggage, while in the second case the carrier cannot claim limitation of liability if the commander was acting within the scope of his employment. [2]

"Negligent pilotage or negligence in the handling of the aircraft, or in navigation" is a comprehensive definition but it does not cover all the errors of the aircraft commander which might lead to an accident. From a remark made during the discussion of Art. 20, it appears that the intention was to make a distinction between "the technical manoeuvres relating to the flying of the aircraft" and "the commercial manipulations not concerned with air law (packing, loading, stowage, etc.)." [3] A similar distinction is made by Van Houtte [4]. Goedhuis points out that this distinction is difficult to maintain, and that incorrect stowage — which may upset the centre of gravity, for example, and cause an aircraft to crash — can also be regarded as an error in the handling of the aircraft [5].

Although we support the objections against differentiation between navigational and commercial errors, we still think that the case of incorrect stowage — like other errors committed before the flight — is not covered by Art. 20.

---

[1] CAB Aircraft Accident Analysis Manual 1949, page 22.
[2] see page 89.
[3] Goedhuis, op. cit. page 234.
[4] cf. van Houtte, op. cit. page 92.
[5] op. cit. page 235.

The original French text referred to "la conduite de l'aéronef", which shows that the criterion to be applied is whether the aircraft was *in movement* or not, but this aspect is lost sight of in the English version, where "la conduite de l'aéronef" is translated as "handling of the aircraft." In our opinion this translation is not quite correct, since it might also include the handling of an aircraft while it is standing on the ground.

It seems reasonable that the dividing line for the application of Art. 20 should be the moment of departure. This will obviate practical difficulties in the interpretation of this article because the moment of departure can be determined with accuracy, whereas the differentiation between flight technical and commercial errors is indeed untenable.

Moreover, this view is wholly in accordance with the spirit of the Convention. Errors occurring in the air, where man is really out of his natural element, may to a certain extent be pardonable and thus exonerate the carrier from liability; to our mind, however, there is not the slightest excuse for extending this principle to errors made before departure, as the flight preparation can be carried out with both feet on the ground, under the all-seeing eye of the operator.

Accurate flight preparation is imperative for safety in air navigation. In this respect ICAO has therefore laid down regulations for flights by transport aircraft [1]. Errors and omissions in the preparation of a flight may lead to accidents, e.g. on taking off without enough fuel or equipment for the proposed flight, or with an aircraft which the commander knew or could have known to be non-airworthy. The cause of accidents of this category is deemed to be "pilot error", but it is not "an error in piloting" and consequently — in our opinion at least — it is not covered by the abovementioned wording of the Warsaw Convention.

If an error in piloting is found to be the cause of an accident which resulted in damage, then the indemnification clause quoted earlier applies as a matter of course; further evaluation of the nature of the error is then unnecessary, unless there is any question of "wilful misconduct or equivalent default."

There has been considerable difference of opinion as to the

---

[1] Annex 6, par. 4.3.

interpretation of that last phrase ever since the **Warsaw Convention** was drafted. If one agrees with Goedhuis [1] that it means a fault so serious as to justify the assumption that the damage was caused deliberately, it will seldom arise as far as the aircraft commander is concerned. Should the court accept this interpretation, then the aircraft commander can generally only be deemed guilty of wilful misconduct, if it is proved that he intended to commit suicide.

A wider interpretation appears to prevail in American case law. The United States Court of Appeals (District of Columbia) has given the following definition: "Wilful misconduct is wilful performance of an act, by the carrier or its employees or agents, with knowledge that the performance of that act was likely to result in injury to a passenger, or performance of an act with reckless and wanton disregard of its probable consequences." [2]

The Warsaw Convention is now being studied by the Legal Committee of ICAO, partly with the object of drafting amendments to remove the uncertainty concerning the interpretation of Art. 25. We merely mention this revision project in passing.

As long as and in so far as the recently revised Rome Convention has not been ratified, liability towards third parties will be determined in accordance with the rules of common law. In the Netherlands it is possible to bring an action for tort in such cases, under Art. 1401 of the Dutch Civil Code [3].

Basing an action on the Rome Convention, however, one must first consider whether the pilot was acting "in the course of his employment."

If the answer is in the affirmative, it does not matter whether he acted "within the scope of his authority" or not. When an aircraft commander arbitrarily deviates from his route and thereby causes damage, he is acting in the course of his employment though not within the scope of his authority. The operator will therefore be liable under the terms of the Convention, but he will be entitled to recourse against the aircraft commander.

The position is different, however, if there is any question of a

[1] Goedhuis, ,,Handboek voor het Luchtrecht", page 259.
[2] cf. ICAO Legal Committee, Minutes and Documents 8th Session, Doc. 7229-I.C/133, page 248.
[3] cf. Sauveplanne, ,,Luchtvaartverzekering", page 78.

"deliberate act or omission ... with intent to cause damage." In that case the operator will only be liable — without any limitation — if it is proved that the employee in question was acting in the course of his employment and *within* the scope of his authority.

In the event of deliberate damage it will usually be impossible to furnish the requisite proof, so that this is a very important escape clause for the operator.

According to common law, the aircraft commander can then be sued direct.

As may be seen from the foregoing, the liability of the aircraft commander will usually be determined in accordance with the rules of common law. In an action for damages the report of the official court of inquiry may serve as a starting-point [1]; an objection to the use of such reports is that in many countries the accident inquiries are handled by the civil aviation authorities. As the latter are also responsible for the functioning of various ground services, this means that in cases where failure of the ground services may have contributed to the accident, the investigating body may actually be inquiring into a case in which they themselves are involved. Holland, with its Air Accident Board, is one of the few countries, where the inquiry into an aircraft accident is held by a completely independent body.

There used to be a similar independent body — the Air Safety Board — in the United States, but since 1940 the accident inquiries in that country have been carried out by the C.A.B., which is also responsible for drawing up standards, rules and regulations concerning matters of safety. However the reports of the C.A.B. cannot be used in civil proceedings [2].

A special system applies in the United Kingdom. There the investigation is generally carried out by the Accidents Investigation Branch of the Ministry of Civil Aviation, but the Minister himself can order an investigation by a special tribunal. As the Minister of Civil Aviation decides whether such an inquiry shall be held or not, and as he also appoints the members of the

---

[1] cf. Guldimann, „Flugunfalluntersuchungen", Schweizer Aero Revue 1951, Nos. 8, 9, 12 and 1952, Nos. 1 and 2.

[2] Cf. Sweeney, "Safety regulations and Accident Investigation", JAL 1950, p. 161 and p. 269; Simpson, "Use of aircraft accident investigation information in actions for damage", JAL 1950, p. 283.

tribunal, it is clear that the accident investigation is not entirely independent of the civil aviation administration [1].

Even though the findings of an inquiry may indicate that an accident was due to pilot error, a civil action for damages brought against the pilot will not necessarily be successful.

Judiciary law on the subject is comparatively scarce.

It is desirable, however, that the court before which a case of this nature is brought, should take into account the exceptional circumstances under which an aircraft commander has to perform his duties, giving due weight to considerations such as those mentioned in the findings concerning an accident to an aircraft which crashed near the Azores in 1949: [2] "The captain apparently had not asked for a radio 'fix' as he would have done had the weather been less favourable, and in the darkness an error in visual estimation had finally arisen, for it is humanly impossible that the judgment of even the most experienced crew member may not be led into error under very exceptional circumstances, when it becomes necessary to make a choice between conflicting data."

---

[1] Cf. detailed "Report on Accident Investigation Procedure", London 1948, Commandpaper 7564.
[2] Interavia Air Letter No. 2011, 1st August 1950, page 3.

CHAPTER IX

THE AIRCRAFT COMMANDER AS EMPLOYEE

LEGAL STATUS OF FLYING PERSONNEL

Except when the aircraft commander is at the same time the owner or operator of the aircraft which he pilots, one of the aspects of his legal status is determined by the fact that he is an employee.

Maritime law contains numerous detailed provisions relating to the master of a ship in his capacity of employee [1].

Foreign air legislation likewise contains special regulations governing the status of flying personnel, including the aircraft commander (e.g. the French [2] and Italian [3] rules for flying personnel, and the German [4] regulations which have meanwhile been suspended). CITEJA also had a series of projects concerning the status of flying personnel on its programme for many years. The latest project, which dates from 1946, comprises provisions on the subject of the contract of employment, compulsory insurance against accidents, and repatriation on termination of employment. [5]

Special rules and regulations dealing with the conditions of employment of flying personnel, however, are not to be found in Dutch law. We must therefore assume that the rules of common law apply in this connection. Accordingly, as the position of the aircraft commander does not differ from that of other employees in this respect, we shall not go into the subject in further detail.

---

[1] Cf. Cleveringa, ,,Het Nieuwe Zeerecht", page 269 onwards.

[2] Le Goff, ,,La Loi du 25 Mars 1936 et le Statut du Personnel navigant de l'aéronautique civile", RGDA 1936, page 145.

[3] Giannini, "Lo Stato Giuridico della Gente dell'Aria".

[4] Döring, ,,Das Arbeitsrecht des Bordpersonals der Deutschen Luftfahrtunternehmen" ArchfLR 1941; Richter, ,,Bemerkungen zum Arbeitsrecht des Luftverkehrspersonals", ArchfLR 1935. For the status of Swiss flying personnel see: Bucher ,,Le Statut juridique du personnel navigant de l'aéronautique civile" and Bratschi, ,,Die Rechtsstellung des Luftfahrtpersonals".

[5] CITEJA Doc. 451.

One cannot help feeling that the special regulations mentioned above are largely based on maritime law rather than on the practical requirements of present-day aviation. The relevant provisions of maritime law frequently have a historical explanation. In the case of long sea voyages on which illiterate hands were sometimes signed on for a single trip, it could be useful to have the conditions of employment legally regulated in detail, preferably on an international basis, in order to prevent abuses. On the other hand, the flying personnel engaged in modern commercial air transportation form a highly select group on the labour market, and as such they have generally proved to be able to negotiate favourable working conditions for themselves.

The short duration of a journey by air reduces the possibility of the crew experiencing undesirable conditions (e.g. poor meals or bad accommodation), and even if such a situation should occur, it is not necessarily of vital importance. In addition, the very high costs attached to advanced instruction and training of flying personnel generally strengthen the ties with the employing companies. Formerly it was quite possible that a seaman might be paid off in some distant port and left to his fate, but this is almost inconceivable in aviation. Statutory regulations concerning a right to repatriation therefore appear to be superfluous. Regulations regarding holidays with pay, health insurance, accident insurance, etc., already appear in the contracts of employment as a matter of course.

For the above reasons, it is not felt necessary to have a special statute governing the legal status of flying personnel in the Netherlands. Common law ensures free consultation between the parties concerned, and in our opinion this is an adequate guarantee of reasonable conditions of employment.

SPECIAL CIRCUMSTANCES OF EMPLOYMENT IN AVIATION

In spite of the foregoing, the circumstances under which the aircraft commander performs his duties present a number of exceptional features. How do these conditions of employment differ from those of the majority of other employees?

The risk element has always characterized the work of the aircraft commander (and of other flying personnel).

In the early years of civil aviation (1921) the expectation of life of a night mail pilot was only 4 years. [1] During the 1922–1925 period, one transport pilot crashed for every 10,000 hours flown in the United States. [2] Notwithstanding technical progress and the accompanying improvement of safety in air navigation, the death rate among flying personnel due to occupational accidents is still higher than in most other branches of industry. During the 1946–1950 period, for example, the accident death rate of airline pilots in the United States was 2.2 per 1.000 pilots per year; the accident figure for the population as a whole was 0.59 per 1,000 inhabitants per year, while the figure for miners was 1.91 per 1,000 per year. [3] One pilot came to the conclusion that "during a 30-year flying career with the scheduled airlines, chances are three in four that a pilot will be involved in an accident, and one in nine that he will be in a fatal accident" [4].

The risk run by military flying personnel is even greater as a general rule; [5] from a study made by the U.S.A.F. School of Aviation Medicine it appears that officers who commence flying at the age of 22 will probably live for 37.6 more years, as against 48.5 years for a comparable group of non-flying officers. [6]

Another characteristic of the pilot's career, and one which is partly connected with the above-mentioned element of risk, is that the period during which this profession may be exercised is relatively short.

In Canada, for example, it has been found that "74 per cent of pilots are through for one reason or another at the age of 34, and after the age of 39 only 9,7% of pilots continue to fly." [7] According to a report published by the British Ministry of Civil Aviation "the age of retirement is about 45 and the pilot is usally compelled to retire because he can no longer comply with the strict medical standard required." [8]

---

[1] Henry Ladd Smith, "Airways, The History of Commercial Aviation in the United States", page 72.

[2] Jerome Lederer, "Safety in the Operation of Air Transportation", quoted by Nicholson, "Air Transportation Management", page 133.

[3] Air Safety Foundation, Accident Prevention Bulletin 52–5, February 21st, 1952.

[4] Captain Moss, "How a Veteran Pilot looks at Safety", American Aviation, February 18th, 1952, page 21.

[5] cf. Sauveplanne, ,,Luchtvaartverzekering", page 15.

[6] Technical Data Digest, 15th May 1949, page 11.

[7] International Labour Organization, Inland Transport Committee, Fourth Session Genoa 1951, General Report I, page 90.

[8] ibidem page 91.

The relatively high occupational risk and the comparatively short productive period create a number of problems of a social nature. It is true that in general the salary level is quite high, but this makes the difference all the greater when the individuals concerned are pensioned or transferred to a ground job while still young; another result of the high salary level is that in many countries it lies above the limit fixed for the general social insurances. Such difficulties can usually be obviated, however, by effecting private insurances or joining pension funds. IFALPA therefore does not regard social security as one of its objectives. [1]

Still another complicated social problem, which we can only mention in passing, relates to the technical progress of aviation and the resultant expansion of productivity, an enormous expansion which is still continuing.

In 1920 a pilot flying a DH-18 with 8 passengers at a speed of 105 m.p.h. produced 840 passenger miles per hour. In 1953 the captain of a Comet with 48 passengers on board and cruising at a speed of 480 m.p.h. produces 23,040 passenger miles per hour.

Even allowing for the greatly increased demand for air transportation, it is not surprising that in view of these figures a fear of unemployment due to technical advances (i.e. "technological unemployment") has been expressed among pilots, particularly in the United States.

In 1945, for example, 4,967 pilots flew an average of 572,519 aircraft miles daily on the domestic network in the United States, whereas in 1948 a total of 4,710 pilots produced a daily average of 925,627 aircraft miles [2]. Moreover, although the era of pilotless aircraft has not yet arrived, this is definitely one of the technical possibilities which must be taken into consideration.

Accordingly, the collective contracts of employment concluded with flying personnel sometimes contain a clause under the terms of which flying personnel who may become redundant as a result of technical progress will enjoy priority in obtaining other functions in the company.

---

[1] ILO report, page 92.
[2] Aviation Week, December 26th, 1949, page 39; cf. also Interavia Airletter No. 2243, 4th July, 1951.

On account of the above-mentioned development, some pilots have expressed a wish to get a larger financial share in this increased productivity, primarily because the faster and larger modern aircraft make heavier demands on them; in 1939 the pilots employed by American Airlines received 0.0075 dollar cents per revenue passenger mile produced, but by 1950 this rate of pay had dropped to 0.0039 dollar cents [1].

In the meantime various American companies have started to base the payment of flying personnel partly on the speed and the size of the aircraft used [2].

In conclusion we wish to mention a problem which also has a special aspect in aviation, viz. the settlement of hours of work and time off. In general, flying personnel have irregular working hours; they perform their duties in widely varying circumstances, and these duties make heavy demands on the physical and psychological powers of resistance of the persons concerned. It is necessary to limit the amount of work which flying personnel may be required to perform, not only for reasons of safety but also for considerations of a social nature. Great difficulty has been experienced in fixing standards for this purpose, as the degree of fatigue depends on a large number of variable factors, such as the type of aircraft, the weather, the navigational aids, the number of intermediate landings, the accommodation on board, the individual concerned, etc. [3]

ICAO has been investigating this problem for several years past but has not yet managed to find a solution. According to the ICAO regulations, [4] the matter is left to the member States on the understanding that each airline concerned must establish limitations on the flight time subject to the approval of the State of registry. Rules of this nature have been drawn up in the Netherlands by the Department of Civil Aviation, acting in consultation with K.L.M. The International Labour Organisation has also

---

[1] F. Lee Moore, "A New Pattern for Flight Pay?", Aviation Week, June 11th 1951, page 55.

[2] cf. Interavia Air Letter No. 2070, 24th October 1950, page 3, idem No. 2235, 22nd June 1951, page 4, idem No. 2243, 4th July 1951.

[3] cf. ICAO Working paper OPS–IV WP17, page 51; detailed list of literature given in "The Role of Fatigue in Pilot Performance", CAA Division of Research, Report No. 61; see also McFarland, "The Human Factor in Air Transport Design"; "Study of Factors contributing to Fatigue in Flight", C.A.A., Washington.

[4] Annex 6, par. 4.2.7.4.

devoted attention to this problem since 1929 and has recommended a study of the matter [1].

THE AIRCRAFT COMMANDER AS A REPRESENTATIVE OF THE OPERATOR

To be able to complete a journey by air quickly and safely, and to acquit himself of the responsibility for the passengers, cargo and mail entrusted to his care, the aircraft commander will sometimes have to perform property transactions. For example, he may have to buy fuel, have repairs made and arrange for meals and accomodation to be supplied to passengers and crew. If, as is often the case, the company has a local agent or representative, the necessary arrangements can be made through the intermediary of that official. But if such an official is not present, as may happen occasionally when scheduled flights are diverted to an alternate airport and will frequently occur in the case of unscheduled flights, the aircraft commander ought to have certain powers in this respect.

As far as the master of a ship is concerned, the whole position is regulated in detail by maritime law; in many cases the master acts as representative of the shipowner [2].

Air legislations of some foreign countries also contain similar provisions relating to the aircraft commander, e.g. France (Law of May 25, 1936), Italy (Law of February 8, 1934) and Uruguay (decree of December 3, 1922) [3].

No such provisions are to be found in Dutch law and most other air legislations.

The carriers have stressed that powers of this nature — if they are really necessary — can be granted to the aircraft commander by power of attorney. But what is the legal position if the carrier has not granted such a power of attorney — as often occurs in actual practice — or if the powers granted prove to be inadequate? In our opinion in those cases the legal security of the parties concerned is not adequately guaranteed by reference to use

---

[1] cf. ,,Généralisation de la réduction de la durée du travail dans le transport par air", RGDA 1938, page 490; ILO Inland Transport Committee, Fourth Session, Genoa 1951, General Report, page 77 onwards and idem Note on the Proceedings, page 46 onwards.

[2] cf. Cleveringa, ,,Het Nieuwe Zeerecht", page 255 onwards.

[3] cf. ICAO Legal Committee, Minutes and Documents 7th Session Doc. 7157/LC 130, page 325.

and wont in connection with deliveries to aircraft. A supplier of spare parts or victuals may find that his bill is not paid by the company, while the aircraft commander who entered into the obligation proves to be insolvent. The commander after returning to his base may be obliged to pay out of his own pocket the costs which he incurred for the purposes of a journey by air. Passengers may experience inconvenience and delay because the aircraft commander refuses to make arrangements on his own responsibility and first requests telegraphic authority from his company.

It therefore appears desirable to regularize the position by statute — if possible, on an international basis — so that the aircraft commander's powers in this respect are established beyond any shadow of doubt. The draft convention on the legal status of the aircraft commander provides for this [1].

---

[1] See page 140.

THE DRAFT CONVENTION ON THE LEGAL
STATUS OF THE AIRCRAFT COMMANDER

CHAPTER I

HISTORY OF THE DRAFT

THE INITIAL PHASE

Even before there was any question of air navigation in the modern sense, legal experts turned their attention to the status of aeronauts. These early studies were usually in the form of a "construction juridique à priori" which was a "pure spéculation savante et subtile" and which would depend on future technical development in aviation for a solid foundation. [1]

It cannot be denied, however, that one of the first studies in this field, an article published in 1891 [2] on "la situation juridique des aéronautes en droit international", takes a practical view of the subject. In this article the author opposed the German view that the French balloonists who made numerous ascents during the Siege of Paris in 1870 — 64 balloons left the besieged city with a total of 155 people on board — must be treated as spies.

These preliminary essays and proposals were largely based on maritime law.

After the Wright brothers had made their first flights in 1903 and Blériot had completed the first international flight by flying across the English Channel in 1909, the year 1919 forms a milestone in the history of aviation and air law. In that year the first air transport companies were founded and the first international air agreement — the Paris Convention — was concluded.

What had seemed a chimera only a few years earlier was now transformed into reality; aircraft carrying passengers took off and flew from one country to another; the speed, range and capacity of aircraft increased by leaps and bounds. The idea that the community on board an aircraft requires leadership, that

---

[1] Maschino, "La Condition Juridique du Personnel Aérien", page 93.
[2] Wilhelm, "De la situation juridique des Aéronautes en droit international", Journal du Droit International Privé 1891, page 440.

there must be a commanding officer on board, was already expressed in the Paris Convention. In Art. 12 of this Convention the "commandant" is the first member of the crew to be mentioned, and the Annexes to the Paris Convention also contain several references to him. On the other hand the Convention gives very little indication of the meaning of this term , the attributes, rights and duties of the aircraft commander.

The Havana Convention (1928) is more explicit, as Art. 25 of that Convention gives the following description of his powers: "So long as a contracting State shall not have established appropriate regulations, the commander of an aircraft shall have rights and duties analogous to those of the captain of a merchant steamer, according to the respective laws of each State".

About the same period, attempts were made to define the status of the aircraft commander in various national legislations: — [3]

Czechoslovakian Law of July 8, 1925:

Art. 16. The navigator of the aircraft, or the pilot, is responsible for the operation of the aircraft and for compliance with the regulations which have to be observed during flight time, from the departure to the landing ... The other operating personnel of the aircraft are subordinate to the navigator or the pilot, as the case may be.

The operating personnel as well as all the other persons taking part in the flight, must obey his orders relating to the necessary discipline in air navigation.

Swedish Royal Decree of April 20, 1928:

Par. 11. The aircraft commander shall be the highest authority on board the aircraft.

Yugoslav Law of February 22, 1928:

Art. 76. The pilot of the aircraft exercises disciplinary power over the crew members and passengers, to the same extent as the captains of seagoing vessels and in conformity with the rules for air traffic.

---

[1] Quoted by Maschino, op. cit. page 97.

Danish Law of May 1, 1923:

Art. 22. On board every aircraft there must be a person who is responsible and who has authority over all the crew and passengers.

The above statutes granted the aircraft commander very wide powers but they were expressed in somewhat vague terms. In addition, they said nothing at all about the aircraft commander's position in private law.

Aeronautical development had meanwhile made this academic question a problem of practical importance. The time therefore seemed ripe for international consultation with the object of reaching agreement on the status of the aircraft commander.

THE STUDIES OF CITEJA

A first International Conference on Private Air Law was convened in 1925 on the initiative of France. At this meeting it was agreed to establish a permanent organization to deal with problems relating to private international air law. The committee set up was called the Comité International Technique d'Experts Juridiques Aériens (CITEJA) and its first session was held in 1926.

Several important conventions have resulted from the work of CITEJA, e.g. the Warsaw Convention (1929), the Rome Convention (1933), the Brussels Convention (1938) and the Convention on Precautionary Arrest (1933). [1]

From the time of its formation in 1926 until its liquidation and incorporation into ICAO in 1947, CITEJA occupied itself with the status of flying personnel in general and the aircraft commander in particular.

The first meeting of CITEJA (May 17–29, 1926) resolved that four Commissions should be created for the study of certain specified subjects. The fourth of these Commissions was charged with the task of investigating the "Condition juridique du Commandant et du Personnel".

Panie, an Italian, was appointed Chairman, and the Belgian

[1] cf. regarding the history and the significance of CITEJA: Riese, "Luftrecht", page 34; Lemoine, "Traité de Droit Aérien" No. 48; Chauveau, 'Droit Aérien", page 335; Goedhuis, "Handboek voor het Luchtrecht", page 8; de Juglart, "Traité élémentaire de Droit Aérien", page 37.

airman-lawyer Thieffry became the rapporteur of the Commission.[1]

It was decided to begin by studying the status of the aircraft commander. Through special circumstances the rapporteur was unable to prepare his report in time for the second session of CITEJA in 1927, so discussion of this subject had to be postponed until the third meeting. [2] At the latter meeting, however, a "Rapport et Avant Project" adopted by the fourth Commission was not discussed [3]. In the interval Thieffry had resigned from the post of rapporteur and he was succeeded by the Pole Babinski [4], who worked out a new draft. At each of the meetings in 1928, 1929, 1930 and 1931, the fourth Commission engaged in long discussions and requested the rapporteur to make alterations in the drafts which he submitted. Finally, in 1931, a draft convention agreed upon by the fourth Commission was discussed at the 6th Meeting of CITEJA, where it was adopted "à titre provisoire."

This reservation was made, not on account of the material contents of the draft but because it was considered desirable to contact the International Labour Organisation (which also had an interest in this project) before formally submitting the draft convention to the French Government.

Moreover, it had been suggested that the draft relating to the aircraft commander ought to be combined with one concerning the crew.[5] Meantime the fourth Commission had begun to study the legal status of flying personnel, a problem into which we shall not go in detail, though it may be mentioned that the solution was made more difficult by the necessary coordination with the ILO, which was dealing with this problem at the same time. [6]

On the outbreak of World War II, CITEJA had produced two drafts relating to closely connected subjects: a draft on the status of the aircraft commander, which had already been adopted "à titre provisoire", and a draft on the status of the crew, for which the preliminary studies were almost finished. The question of

---

[1] Compte Rendu de la 1ère Session, page 50 and page 39.
[2] Compte Rendu de la 2e Session, page 13.
[3] Compte Rendu de la 3e Session, page 76.
[4] Compte Rendu de la 4e Session, page 10.
[5] Compte Rendu de la 6e Session, page 72 and Resolution No. 47, page 170.
[6] Compte Rendu de la 7e Session, page 14.
   Compte Rendu de la 8e Session, page 33.
   Compte Rendu de la 10e Session, page 15.
   Compte Rendu de la 11e Session, page 15.
   Compte Rendu de la 12e Session, pages 43 and 81.

combining the two drafts was raised at the last meeting prior to the War — the 13th — but no decision was reached. [1]

The first post-war meeting of CITEJA took place in 1946. The Frenchman Garnault succeeded Babinski as rapporteur and it was decided (a) that the draft relating to the aircraft commander should be revised in the light of recent technical development and (b) that this draft should be combined with that relating to the crew. [2] At the next meeting of the fourth Commission, which was held in Paris during July, 1946, it was found that the American and British delegates had serious objections against combination of the two drafts. [3] In view of this it was resolved that the status of the aircraft commander should again be treated separately at the session of CITEJA to be held in 1947.

At the 1947 session, i.e. the 15th session, which was held in Cairo, CITEJA adopted a draft convention and a resolution worded as follows: [4]

"Le CITEJA ...

Charge son Secrétaire Général de transmettre aux Etats adhérents au Comité, et à l'O.P.A.C.I., le Projet de Convention Internationale relatif au Statut Juridique du Commandant d'Aéronef adopté lors de la présente Session;

Emet le Voeu que ledit Projet soit soumis à l'approbation d'une Conférence de Droit International Privé Aérien convoquée par les soins de l'O.P.A.C.I. ..."

ICAO AND THE DRAFT CONVENTION

The possibility of incorporating CITEJA into CINA had been contemplated even before the War, but at that time it was objected that CITEJA included experts from countries which did not belong to CINA. This objection did not apply in the same measure to (P)ICAO, which was to have a more universal character [5].

At the Chicago Conference in December, 1944, the post-war status of CITEJA came up for discussion and a resolution was

---

[1] Compte Rendu de la 13e Session, page 20.
[2] Compte Rendu de la 14e Session, page 96.
[3] Rapport et Avant-Projet de Convention par M. Garnault, Doc. No. 451, page 2.
[4] Compte Rendu de la 15e Session, Resolution No. 161, page 92.
[5] cf. Latchford, ,,Coordination of CITEJA with the new International Civil Aviation Organizations", Department of State Bulletin, February 25th, 1945, page 310.

adopted to the effect that (1) resumption of the activities of CITEJA was desirable and (2) that consideration should be given to possible ways of securing coordination with ICAO [1].

At the first Interim Assembly of PICAO it was resolved that a Legal Committee should be established within the framework of ICAO and that, in consultation with CITEJA, the functions of the latter body should be taken over by this Legal Committee. [2] CITEJA's decision to hand over the draft convention on the legal status of the aircraft commander to ICAO preceded the ultimate liquidation of CITEJA.

At the 16th and final session of CITEJA, which took place in Montreal during May, 1947, this body was liquidated and the archives were transferred to the Legal Committee of ICAO. [3]

After the text of the CITEJA draft had been revised by a Legal ad hoc Committee of PICAO, the project was put on the agenda for the first General Assembly of ICAO.

It was not discussed, however, and the Assembly resolved to place it on the working programme of the Legal Committee. [4]

When the Legal Committee met for the second time in June, 1948, it was decided that a first essential was to learn what technicians thought of the draft. For this purpose the draft was transmitted to the Council with a request for comments from the technical bodies of ICAO. [5]

At its seventh meeting, in January, 1951, the Legal Committee asked the Council's views on:

"I. the need for a convention on the legal status of the aircraft commander;

II. the technical or economic aspects of the problem." [6]

On completion of this study, opinions had been received from some sections of the permanent ICAO Secretariat, viz. the Air Transport Bureau, the Air Navigation Bureau and the Legal Bureau.

---

[1] cf. Latchford, "Private International Air Law", Department of State Bulletin, January 7th, 1945, page 28.

[2] cf. Latchford, "Private International Air Law Developments", Department of State Bulletin, November 17th, 1946, page 883.

[3] Compte Rendu de la 16e Session, page 36 onwards.

[4] Minutes and Documents, Legal Committee First Session, page 208.

[5] Minutes and Documents, Legal Committee Second Session, page 12.

[6] Minutes and Documents, Legal Committee Seventh Session; Doc.C-WP/980 page 1.

Comments were also received from IATA, IFALPA and the United Kingdom [1]. Guatemala expressed her agreement with the draft. The Council, however, had not yet decided what further action should be taken with the draft.

THE DESIRABILITY OF A CONVENTION

As may be seen from the foregoing, the drafting of an international convention on the legal status of the aircraft commander has figured on the working programme of CITEJA, and later of the ICAO Legal Comittee, for 25 years without any definite result. This is all the more remarkable when one considers that the character of aviation is such that urgent problems ought to be solved without undue delay.

Before drawing the obvious conclusion from this slow progress, that a convention is unnecessary — or at least not particularly urgent — it may be useful to analyse the causes of the delay.

To begin with, it must be pointed out that although the first rapporteur had to terminate his work prematurely, which undoubtedly caused some delay because his successor had to familiarize himself with the subject, a draft convention was adopted — though only "à titre provisoire" — within five years. In this connection it should be remembered that CITEJA was in a difficult position; officially it could only deal with questions of private law, and yet it was soon found that the status of the aircraft commander is also a matter of public law.

Unfortunately, after this temporary success CITEJA decided to combine the status of the aircraft commander with the regulations concerning the conditions of employment of flying personnel in general, a controversial subject on which agreement has not yet been reached. This decision may have been formally correct, but it meant that the draft relating to the aircraft commander was left lying for many years.

Next came the unavoidable interruption due to World War II, at the end of which the draft had to be re-adapted to the altered circumstances then prevailing, while the status of CITEJA itself was also uncertain for some time.

To sum up, delay was more or less inevitable during the period

---

[1] cf. Doc. C–WP/980 and Doc. C–WP/899.

when CITEJA was dealing with the question. But the same cannot be said of the period following 1947, when what was really a completed project — representing the outcome of countless discussions — had been handed over to ICAO by CITEJA.

It might have been expected that the question in abeyance could at last be settled definitely, especially since ICAO can also deal with problems of public law and is therefore more favourably placed than CITEJA in this respect. ICAO, however, has made very little progress with the project. In our opinion this disappointing result is solely due to the fact that the Legal Committee has numerous problems on its agenda and the question of the status of the aircraft commander has not been given high priority. The opposition of IATA to the project may perhaps have been partly responsible for this.

As far as the necessity for a convention is concerned, there does not seem to be much difference of opinion among those who have devoted attention to the subject.

If one regards an aircraft in flight as a small community temporarily isolated from the rest of the world, it is obvious that this community — with all its members — requires a jurisdiction with a legal foundation; the international character of aviation makes it advisable that regulations of this nature should be internationally accepted.

More than thirty national legislations contain provisions relating to the status of the aircraft commander. On the one hand this shows that the necessity of such regulations is generally recognised, but on the other hand the diversity of these regulations accentuates the need for uniformity.

Most European countries have displayed interest in the project, and not a single country has questioned the desirability of a convention. The American CITEJA experts were strongly in favour of it, while the representatives of the American Airline Pilots Association and the International Federation of Airline Pilots Associations have likewise advocated a convention of this nature. [1]

The same may be said of many eminent lawyers who have studied the subject within the framework of CITEJA or ICAO.

---

[1] cf. Doc. C–WP/980, page 2; Doc. C–WP/899, pages 2 and 3, and Knauth, "The Aircraftcommander in International Law", JAL 1947, page 160.

Last but not least, many authors have referred in various ways to the desirability of a convention on the legal status of the aircraft commander. [1]

It has been suggested that the matter might be settled on an international basis by means of an Annex to the Chicago Convention, [2] but in our opinion the drawback of this is that an Annex does not offer any guarantee of uniformity, because the Chicago Convention allows the States to deviate from an Annex. We therefore prefer that the position of the aircraft commander should be regularized by means of a convention, which would also emphasize the importance of the principles involved.

As an exception to the above-mentioned unanimity, on several occasions IATA has indicated that it considers a convention to be premature. One may observe, however, that the objections formulated by IATA were not always of a purely legal character. At the 29th meeting of IATA in 1933, for example, the rapporteur moved that CITEJA be requested to postpone the work of drafting a convention on flying personnel " . . . rather because of practical than legal considerations. Traffic and considerations of economy in general cannot stand the financial and technical burdens which are contemplated in this connection until a definite necessity for such proposals becomes apparent. Such necessity does not exist". [3]

This aroused the following criticism from a delegate: "I should like to know why anyone wants to hold up other people's work? What is to be gained by such an attitude? It is necessary, of course, to consider what the financial burdens would be, but the members of IATA would do more useful work if they were to state what matters of general interest they think ought to be dealt with in a draft convention on the position of the navigating crew, rather than place difficulties in the way of the preliminary work". [4]

---

[1] cf. Knauth, loc. cit.; Honig, "De Positie van de Gezagvoerder van een Luchtvaartuig", NJB 1951, page 317; Bucher, "Le Statut juridique du Personnel navigant de l'aéronautique", page 70; Charlier, "Le Commandant d'aéronef en droit privé", RGDA 1947, page 21; Babinski, "l'Aspect juridique de la notion du Commandant de l'aéronef", Riv. Dir. Aer. 1932; Chauveau, „Droit Aérien", page 453. Contra: Shawcross and Beaumont, "Airlaw", page 487.

[2] By the United Kingdom and the Air Transport Bureau of the ICAO, cf. Doc. C–WP/980, page 5 and C–WP/899, appendix C.

[3] Döring in IATA Information Bulletin No. 19, page 8.

[4] Plesman, ibidem page 10.

In 1947 IATA again affirmed in a letter to ICAO that "there is no immediate need for a convention of this character." [1]

It is important to remember that IATA only represents the airlines which operate scheduled services. The problem of the status of the aircraft commander is naturally less pressing for "multi-million dollar corporations sponsored by States as a matter of policy", [2] as they have an efficient ground organization everywhere. But, as remarked by the Legal Bureau of ICAO, one must also think of the smaller companies and free-lance pilots: " ... uncertainty (of status) may cause no harm in the case of commanders of aircrafts belonging to regular and well-known airlines, but may present difficulties in the case of non-regular airlines or small operators, some of whom may be, at the same time, their own aircraft commander." [3]

Chauveau expresses the view that the slow progress of the project is the result of " ... des observations réticentes, mais croit-on mal fondées, de l'IATA." [4]

Knauth justifiably remarks that if one were to wait until the problem becomes acute, it might be too late. Moreover, it is easier to reach agreement if little or no precedent exists.

We shall end our consideration of the need for such a convention by quoting Knauth's conclusion, viz. "The commander's status is today quite undefined. Surely a man employed to exercise some sort of control over a half million dollar machine, with a crew of 5 to 10 persons, with 30 to 80 passengers, cargo and mails of high value, and travelling over and to many jurisdictions should certainly have his status, rights and powers carefully and clearly stated in every language. The public interest demands it." [5]

EVOLUTION OF THE ICAO DRAFT

Numerous drafts have been prepared within the framework of CITEJA and ICAO since the time when CITEJA first began to occupy itself with the status of the aircraft commander: the draft of Thieffry (1927), the drafts of Babinski (1930 and 1931), the

---

[1] ICAO Doc. C–WP/899, Appendix D.
[2] C–WP/980, pag. 7.
[3] ibidem, page 9.
[4] Chauveau, "Droit Aérien", page 454.
[5] Knauth, JAL 1947, page 160.

drafts of Garnault (1946) and finally the revised ICAO draft (1947). We can hardly study the contents of all these proposals in detail. It may be of interest to mention some of the main trends in the changing line of thought which they disclose, however, because they reflect the development of aviation during the past quarter of a century.

The original intention was to regularize the capacity of the aircraft commander in private law, his legal and commercial power to act as representative of the "avioneur". He was regarded as the leader of an "expédition aérienne," who required extensive powers to ensure the success of such an enterprise — since forced landings in remote regions were then an everyday occurrence.

As air transportation became more of a routine operation, as the network of telecommunication facilities was extended, and as the large airlines began to station agents and representatives all along the routes, the private law element receded into the background. In the course of the discussions it soon became apparent that the public law element could not be ignored. For this reason Babinski's first draft dealt with the following aspects: "institution du commandant et formalités qui y sont attachées, pouvoirs de discipline et de hiérarchie, commandant comme mandataire du propriétaire, commandant comme mandataire des chargeurs, responsabilité, commandant comme officier public." [1] Except for the rules concerning liability, these aspects are still to be found in the present draft. The controversy over the problem of liability — described as "la question la plus difficile" [2] — proved to be so great that finally the subject was left out of the Convention altogether.

Initially the aircraft commander was thought of as an extra member of the crew, as a sort of supernumerary.

Although some people at once envisaged the possibility of combining the function of commander with that of navigator or pilot, objections were raised against such a pluralism. It was pointed out that the general education of the pilots often left

---

[1] Babinski in Rapport sur la situation juridique du Commandant, Doc. 17, page 3.
[2] Babinski in Compte Rendu des Réunions de la 4ème Commission, May 1931, Doc. 84, page 45.

something to be desired, and objections were also made on grounds of safety:

" ... il arrivera que dans un cas difficile, dans une situation critique, le pilote sera obligé de manoeuvrer rapidement, et ses manoeuvres seront gênées par les responsabilités qu'il aura. Dans ces cas, il importe qu'il y ait un commandant; il y a intérêt à ce que celui qui est responsable de la manoeuvre ne soit pas celui qui l'exécute. Ceci est tout à fait important. Il faut que le pilote soit une machine à qui on dit: Faites ceci! Faites cela!, sans qu'il ait la responsabilité de la manoeuvre". [1]

The above concept of the aircraft pilot, which likens him to the helmsman rather than to the master of a ship, is somewhat different from the now generally accepted idea of a pilot-in-command.

The original idea of the commander being a supernumerary was connected with the initial proposal that the appointment should be optional; this was because of a reluctance to impose unnecessary obligations on aviation, which was faced with great difficulties in this early period. On the other hand the facultative nature of the obligation to designate an aircraft commander seemed to make the usefulness of a convention somewhat illusory.

Accordingly, the next step was to find criteria for categories of aircraft on which it would be obligatory to have a commander on board. After beginning with transport planes, in general, the requirement was later confined to transport planes of a certain size, while the distance to be covered was also suggested as a criterion. The scope of the draft convention was subsequently extended to include all international commercial flights and finally all international flights. It has already been proposed that the latter restriction should likewise be dropped and the appointment of an aircraft commander made obligatory for *all* flights, i.e. including non-international flights. [2]

A change of attitude is also perceptible in the manner of regulating the powers of the commander in private law. In Thieffry's draft it was left entirely to the operator to state the extent of the commander's powers in the letter of appointment [3]. Babinski's

---

[1] Vivent in Compte Rendu des Réunions de la 4ème Commission, May 1930, Doc. 37, page 11.

[2] Charlier, "Le Commandant d'aéronef en droit privé" RGDA 1947, page 21.

[3] Article 5 in Avant-Projet Thieffry (1927).

drafts introduced an intermediate form, for although he listed a number of statutory powers, these could be restricted or extended by the operator [1]. Lastly, the ICAO draft gives a limitative summary of the commander's powers; these powers are mandatory and the operator cannot curtail them as far as third parties are concerned.

In the following chapter we shall study the various articles of the latest version (the ICAO draft) of the convention on the legal status of the aircraft commander, referring from time to time to the earlier drafts and the relevant discussions.

---

[1] Art. 8 and Art. 10 in Doc. 17.

CHAPTER II

COMMENTARY ON THE ARTICLES OF THE DRAFT CONVENTION

ARTICLE 1

(1) Every aircraft performing an international flight shall carry one person vested with the powers of a Commander.

(2) The right to designate the Commander belongs to the operator of the aircraft.

(3) In the absence of any Commander so designated, or in case the latter is prevented from performing his duties, and if no successor has been designated by the operator, the Commander's duties will be carried out by the other members of the crew in the following order: pilots, navigators, engineers, radio operators and stewards. The order of succession within each category shall be determined in accordance with the rank assigned by the operator.

This first article contains a number of terms which are not defined anywhere in the Convention, viz. "international flight", "Commander" and "operator".

The desirability of giving definitions has been stressed in various quarters during discussions on the Convention, and the task was indeed attempted in some of the earlier versions.

One definition suggested for the term "international flight" was: "un vol dans le territoire d'un autre Etat contractant." [1] Garnault's draft included the following definition: "un aéronef accomplit un vol international, lorsqu'il survole le territoire de deux ou plusieurs états." [2] It has also been recommended that the Warsaw Convention definition should be retained.

The commander, originally referred to as the "capitaine," [3]

---

[1] Observations présentées par le délégué Norvégien, Doc. 415, page 2.

[2] Rapport et Avant-projet de Convention par M. Garnault, Doc. 434, art. 1.

[3] Avant-Projet Thieffry (1927), article 1.

was often described by enumerating some of his powers: "Le commandant de l'aéronef est la personne investie de pouvoirs d'autorité et de discipline à bord d'un aéronef et représentant le propriétaire et les chargeurs" [1]. The definition proposed by Garnault was: "le commandant de bord est le chef de l'équipage à bord de l'aéronef en vol", later revised as follows: "le commandant est le membre du personnel navigant qui est le chef à bord de l'aéronef." [2]

As far as the operator is concerned, in the first draft it was proposed that the term "avioneur" should be introduced, but this term was not definied. In the absence of the "avioneur", his rights devolved upon the owner. [3]

Babinski proposed an article reading as follows: "Partout dans la présente Convention où il est question du propriétaire de l'aéronef, on lui assimile l'exploitant, lorsque le propriétaire a cedé l'exploitation à une tierce personne, qui effectue les transports en son propre nom." [4] It was unanimously agreed to delete this article, however, and to substitute "l'exploitant" in place of "le propriétaire" throughout the draft [5]. Nevertheless the latter term was not defined. In a later draft the rapporteur made the following comment: "La Convention ne s'occupe pas de la question de savoir, quel est le lien juridique entre l'appareil et l'exploitant. Il peut être propriétaire, locataire, usufruitier, etc., ce qui importe pour les besoins de la Convention, c'est le fait d'exploiter l'aéronef". [6]

The problem of the definition of "operator" again came up for discussion after the War, [7] but a proposal that the term should be given the same definition as in the Rome Convention was rejected. [8]

As may be seen from the foregoing, this question has received a great deal of attention and yet the draft still does not contain any definitions. The view which finally prevailed was that, for the sake of uniformity in air law, it would be undesirable to include

---

[1] Rapport par M. Babinski, Doc. 17, article 1.
[2] Rapport et Avant-projet de convention par M. Garnault, Doc. 434, article 1.
[3] Avant-projet Thieffry (1927), article 1.
[4] Rapport par M. Babinski, Doc. 17, art. 19.
[5] Compte Rendu des Réunions de la 4ème Commission, Doc. 37, page 60.
[6] Avant-Projet de Convention par M. Babinski, Doc. 119, page 7.
[7] Compte Rendu des Réunions de la 4ème Commission, Doc. 493, page 54.
[8] Compte Rendu des Réunions de la 4éme Commission, Doc. 496, page 7 onwards.

special definitions which might differ from those appearing in other Conventions.

On the other hand, however, the authors of the Convention did not want to tie themselves down to definitions already used elsewhere. In the end it was decided that the drafting of the definitions should be left to (P)ICAO, under the terms of the following resolution:

"Le CITEJA ... souligne l'intérêt qui s'attache à ce que les expressions: exploitant, transport international, employées dans le Projet, reçoivent de l'OPACI une définition en accord avec le sens donné aux dites expressions dans les diverses Conventions Internationales de Droit Aérien actuellement en vigeur." [1]

As far as the terms "operator" and "international flight"are concerned, it will therefore be necessary to await the decision of ICAO.

Although the "aircraft commander" is not defined either, Art. 1 (1) indicates which person is referred to, viz. the person "vested with the *powers* of a Commander."

The powers in question are set out in the later articles of the Convention. Here we are at once faced with an anomaly, for the aircraft commander has not only powers, but also — and perhaps most important of all — duties. Some of these duties are mentioned in the draft — see Arts. 2 (*b*) and 7. It would therefore be more correct if the first paragraph were to speak of: " ... one person vested with the powers and *charged with the duties of a commander.*"

As remarked earlier, [2] the obligation to carry a commander on every international flight is the (provisional) outcome of a long process of development. It has already been suggested — and rightly so, in our opinion — that the rules of this Convention should also be applicable to non-international flights: "Nous souhaiterions même qu'elles soient obligatoires aussi pour la navigation purement nationale, qui sans cesse rencontre celle internationale, la croise, la double, partage avec elle le trafic, lui succède ou la précède, si bien qu'elles forment une ensemble dont le régime doit être unifié". [3] One might add that in some countries of great extent (e.g. the interior of Australia and certain

---

[1] Resolution No. 161 in Doc. 485, page 92.
[2] See page 128.
[3] Charlier, "Le Commandant d'Aéronef en Droit Privé", RGDA 1947, page 21; similarly Georgiades in Doc. 432, page 6.

parts of South America) there may be more need for an aircraft commander with clearly defined rights and duties on domestic routes than on the international routes of Western Europe. The value of the draft Convention would be increased if aircraft commanders on non-international flights were to be given the status provided for international flights in the present draft. This could easily be done by altering the first paragraph to read " ... a flight ..." instead of " ... an international flight ...".

Another point to be noted is that the first paragraph of Art. 1 makes it obligatory to have a commander on board the aircraft; in other words, flights by pilotless aircraft without persons on board are forbidden. This absolute prohibition goes much further than the Chicago Convention, in Art. 8 of which it is prescribed that pilotless aircraft may only be flown over the territory of the contracting States if permission has been obtained and if the flights are performed in such a way that other aircraft will not be endangered.

As far as could be ascertained, attention has never been drawn to the above consequence of the draft Convention, but it is possible that some countries may object to ratification on this ground.

The appointment of a commander is based on the principle that the miniature community on board an aircraft requires the leadership of a person vested with statutory authority. Naturally this consideration applies exclusively to flights in which there are persons on board. In the future, if aircraft are controlled from the ground, it may become necessary to make a distinction between flights with persons on board and flights with freight and mail only. In the first case there will have to be an aircraft commander on board, but in the second case this may not be essential. To forestall possible objections, it might therefore be advisable to amend the first paragraph of Art. 1, so that the obligation at present embodied in it will only exist if people —either crew or passengers — are on board.

Par. 2 recognizes the operator's right to designate the commander of the aircraft, but this is not an obligation.[1] How must the commander be designated, if he is designated at all? The draft is

---

[1] Otherwise: Charlier, loc. cit. page 23.

silent on that point. In the first draft there was some question of a "letter of appointment," [1] and according to later drafts the appointment had to be recorded in the log books. [2] It is regrettable that the latest draft is not clear on this subject. For lack of a definite ruling, one must assume that the manner of appointment is not limited in any way; the commander can be designated either verbally or in writing, and an entry in the log books is unnecessary, according to the draft at least.

Otherwise the point is not of much practical importance, because the 1st pilot automatically acts as commander, under the terms of Par. 3, if no commander has been designated. This member of the crew will practically always be selected by the operator to act as commander of the aircraft. Accordingly, whether the operator designates him or not, and irrespective of the manner in which this is done, the position really remains the same.

The rules in Par. 3 concerning the order of succession, which come into effect in the absence of a designated commander, or if the latter is prevented from performing his duties, are a comparatively recent innovation. These rules were inserted in the draft at the suggestion of American delegates, with the object of ensuring that even under unforeseen circumstances, e.g. in the event of an accident, there would always be somebody present vested with the statutory powers of an aircraft commander [3]. Although some people wondered whether an international convention was really the appropriate place for detailed rules of this nature, and there was also an objection to the possibility of far-reaching powers being granted to "des employés tout à fait subalternes" [4], the American proposal was adopted and later included in the ICAO draft. We have no serious objection to these rules but it is doubtful if they have any real use.

One might also wonder whether the stewardesses come under the category of "stewards", or whether they were intentionally omitted from the order of succession.

A more important point is that — probably through carelessness — the real aim is not achieved through the present wording.

[1] Avant-Projet Thieffry (1927), article 5.
[2] Avant-Projet de Convention par M. Babinski, Doc. 119, art. 2.
[3] M. Knauth in Compte Rendu des Réunions de la 4ème Commission, Doc. 493, page 42.
[4] M. Garnault, ibidem page 59.

According to this draft, the successor merely falls heir to the *duties* of the commander; but the intention was that he should take over the *rights* of the aircraft commander. In our opinion it would therefore be better to say: " ... if no successor has been designated by the operator, one of the other members of the crew will act as Commander in the following order ..."

The next question to be considered is what happens — from a legal aspect — if one of the persons on board usurps the function of commander. Such a thing may well happen (it will be recalled that several instants of this have been reported in the newspapers in recent years).

Here one must make a distinction between the following two alternatives:

a. One of the passengers may overpower the commander and take command of the aircraft. Quite apart from the possibility of such an act being a criminal offence, in these circumstances the usurper is not entitled to assume either rights or duties, because only a member of the crew is legally entitled to act as the commander's successor.

b. On the other hand, one of the members of the crew, e.g. the co-pilot, may tie up the commander and take over the command. In some quarters it is held that the crew member in question then becomes the legal successor of the aircraft commander. Charlier, for example, remarks that: "du moment où, à bord, il y a un Commandant de fait et où aucun Commandant légal ne lui reprend la direction, l'équipage et les passagers doivent lui obéir si cela est nécessaire pour tout ce qui concerne la sécurité et la police" [1].

It is difficult to accept such a construction based on *de facto* authority. Furthermore, the reasoning is self-contradictory, since it is stated that the persons on board must obey the new commander and yet it is implicitly acknowledged that the deposed commander may re-assume command.

ARTICLE 2

(1) Within the periods specified in Article 5 below, the aircraft commander:

---

[1] Charlier, loc. cit. page 25.

a) *shall be in charge of the aircraft, the crew, the passengers, and the cargo;*

b) *has the right and the duty to control and direct the crew and the passengers to the full extent necessary to ensure order and safety;*

c) *has the right, for good reason, to disembark any number of the crew, or passengers at an intermediate stop;*

d) *has disciplinary power over members of the crew within the scope of their duties; in case of necessity, of which he shall be sole judge, he may assign temporarily any member of the crew to duties other than those for which he is engaged.*

This article deals with the commander of an aircraft in the literal sense of "commander". According to the English text he is "in charge" of the aircraft, the crew, the passengers and the cargo. The intention is clearer in the French text [1], which says that he "exerce le commandement".

The aircraft commander is thus granted a status closely analogous to that of the master of a ship, whom the law makes an "uncrowned king" on board his ship [2].

Like the master of a ship, however, the aircraft commander is bound to exercise the wide authority conferred upon him for the purposes for which it was intended; it must not serve to satisfy a personal craving for power. With this in view, certain restrictions are imposed on him with regard to the use of his authority, though the limitations are not defined in the same manner for all of the powers granted to him.

The power granted under *b*, for example, may only be exercised "to ensure order and safety," whereas the power mentioned under *c* can be used "for good reason." Again, the disciplinary power over the crew is only "within the scope of their duties", and the other power mentioned under *d* can only be exercised "in case of necessity, of which he shall be sole judge." Through all these limitations the authority of the aircraft commander is kept within its natural bounds.

Art. 2 distinguishes between crew and passengers. The category to which the persons on board belong will have to be

---

[1] The French, English and Spanish texts are equivalent in accordance with article 11 of the draft convention.

[2] Cleveringa, ,,Het Nieuwe Zeerecht'', page 211.

established by reference to the log book (in which the names and functions of the crew members must be entered) and also from the passenger manifest (which must likewise be carried on board the aircraft), though the draft convention makes no mention of this.

The distinction is important, because the commander does not exercise the same authority over both categories. Paragraphs *a*, *b* and *c* apply to both groups, but besides this under the terms of Par. *d* the commander exercises disciplinary power over the crew.

An apparent defect of the present wording is that a stowaway is neither a member of the crew nor a passenger, and consequently the commander would really have no jurisdiction over him. It might therefore be better to amend the wording in such a way that Pars. *a*, *b* and *c* refer to "all persons on board".

A second shortcoming, in our opinion, is that according to Par. *c* the aircraft commander can disembark passengers and crew, but there is no mention of the cargo. Cases may occur where the aircraft commander finds himself obliged "for good reason" to have cargo unloaded at an intermediate stop. Some of the cargo, for example, may prove to be troublesome or even dangerous for the persons on board; when the cargo includes livestock or perishable goods, it is quite conceivable that their condition may necessitate disembarkation. It appears to be desirable that this power, which is exercised by the commander in actual practice and seems to be generally recognised, should be given legal foundation. Par. *c* ought to be amended accordingly.

It has been proposed that the aircraft commander should be granted the right to dispose of the cargo if he considers this necessary for the preservation of the aircraft or the persons on board, in analogy with the provisions of maritime law. In this connection one need only think of the jettisoning of cargo when the aircraft gets into difficulties while in flight, or the consumption of any victuals which happen to be available after a forced landing. We think that such an amplification is superfluous. Ignoring the fact that the situations in question will very seldom arise and that the adage "necessity knows no law" may apply in such a case, the aircraft commander is already in charge of the cargo under the terms of Par. *a*. To our mind it would not be an excessively wide interpretation if the aircraft commander were to derive from this clause the right to dispose of the cargo in case of emergency.

We have already mentioned that the commander has more authority over the crew than over the other persons on board. The commander has "disciplinary power" over the members of the crew. What does this involve? The expression appeared in the very first draft of Babinski, and in Art. 5 it was expressly stated that this included the power to impose disciplinary punishments as well as the power to give orders, "qui doivent être exécutés intégralement." [1] The question of the penalties to be imposed received special attention, not so much because it was desired to withold this power from the commander but because of the doubt as to which law would be applicable.

It was pointed out that in some countries corporal punishment may be imposed, whereas in other countries this is barred. [2] The consensus of opinion was that the "law of the flag" (in other words the law of the state of registry) ought to apply, but the English delegates were opposed to this. [3] In the first draft it was stated that the commander could impose the disciplinary punishments "prévues dans les règlements de service", which included the regulations laid down by the authorities as well as those of the airlines.

In a later text this provision was extended to "ayant force de loi." [4] On the one hand this addition gave expression to the idea that the regulations ought to have a legal basis, but on the other hand it did not settle the question of which law should be applicable. The German delegate considered such a solution unsatisfactory; he suggested a more precise ruling on the question of jurisdiction, but as he himself remarked: "en introduisant la loi nationale de l'aéronef, la loi du pays survolé et l'ordre public, on n'arrive pas à une unification du droit. C'est pour cela qu'on propose de ne pas mentionner expressément le droit d'infliger des peines, mais de se borner à la constatation que le commandant a le droit de prendre les mesures appropriées pour assurer l'observation de ses ordres." [5] The object of this change in the wording was to avoid jurisdictional difficulties, but it was definitely not

---

[1] Rapport par M. Babinski, Doc. 17, article 5.
[2] Babinski in Compte Rendu des Réunions de la 4 ème Commission, Doc. 37, page 18.
[3] ibidem page 19.
[4] Rapport supplémentaire par M. Babinski, Doc. 67, article 4.
[5] Compte Rendu des Réunions de la 4ème Commission, Doc. 84, page 22.

the intention to limit the powers of the commander, as is clearly evident from the subsequent discussion:

"M. YOUPIS (Greece) — Le texte dirait: "Les pouvoirs de discipline envers le personnel navigant (équipage) consistent en la faculté de leur donner des ordres qui doivent être exécutés intégralement et d'assurer l'observation des règlements de service".

Sir Alfred DENNIS (Great Britain) — Alors, il n'a pas le pouvoir d'infliger des peines?

M. BABINSKI, Rapporteur — Dans la proposition de M. Youpis il n'est pas question de peines.

M. GARACHANINE (Yugoslavia) — Quelle sera la sanction, s'il ne peut pas infliger des peines?

M. Youpis (Greece) — S'il y a des peines prévues dans les règlements, le commandant les infligera, mais pour qu'il puisse les infliger, il faut qu'elles soient permises par la loi du pays, de telle sorte que l'ordre public du pays ne soit pas en contradiction avec l'application du règlement. Cela se comprend.

M. RIESE (Germany) — Le commandant peut infliger des peines si elles sont prévues dans le règlement de service, et il va de soi que ce règlement de service ne peut pas prévoir des peines contraires à la loi.

Sir Alfred DENNIS (Great Britain) est d'accord dans ces conditions." [1]

In spite of this, at the 6th meeting of CITEJA, the Italian delegate Ambrosini proposed that every definition of the authority of the commander, not only over the passengers but also over the crew, should be omitted, for the following reason: "Il vaut mieux laisser au texte un caractère général, car il est impossible de déterminer quels seront les pouvoirs du commandant, en cours de vol. Il faut que le commandant de l'aéronef ait tous les pouvoirs, sans aucune limitation .... Il propose la suppression des articles 4 et 5. Il propose surtout la suppression de l'article 4, car il trouve dangereux de définir les pouvoirs du commandant vis-à-vis de l'équipage quand l'aéronef est en vol. A ce moment-là, le commandant doit avoir pleins pouvoirs vis-à-vis de l'équipage, il doit pouvoir lui donner les ordres qu'il juge utiles, même les ordres les plus onéreux." [2]

---

[1] Compte Rendu des Réunions de la 4ème Commission, Doc. 84, pages 23 and 24.
[2] Compte Rendu de la 6ème Session, page 84.

Ambrosini's proposal was adopted by nine votes to five, which shows that the majority concurred with this view.

When study of the project was resumed after the War, some special attributes were again added to the summary statement of the commander's authority on board the aircraft, though these attributes may really be taken for granted, e.g. the clauses relating to disembarkation and assignment of duties.

For the interpretation of the scope of the commander's authority, however, the previous history of the draft is still important.

ARTICLE 3.

(*1*) *The aircraft Commander shall have the right, without special authority*:

a) *to buy any items necessary for the completion of the trip*;
b) *to have any repairs made which are necessary to enable the aircraft to proceed promptly on its trip*;
c) *to make any arrangements and to undertake any expenditure which may be necessary for securing the safety of the passengers and crew and the preservation of the cargo*;
d) *to borrow the sums required for the accomplishment of the measures mentioned in Pars. a, b and c of this article*;
e) *to engage, for the duration of the trip, in replacement of members of the crew who cease to be available for any reason, such personnel as is essential for the completion of the trip*.

The difficulties in reaching agreement on a convention relating to the status of the aircraft commander are in no small measure due to the controversy over the subject matter of this article.

One of the essential points even in the first draft was the intention to make the aircraft commander the representative, in a certain sense, of his employer, and this particular proposal has encountered very strong opposition from IATA. The objections raised are mostly in connection with the following points.

A. The personnel to be entrusted with the function of commander often lack the necessary general education for them to be given such far-reaching powers.

Babinski made the following comment about the function of pilot being combined with that of aircraft commander: "Si les fonctions du commandant sont très larges et nécessitent des connaissances spéciales dans le domaine administratif, juridique et commercial, la séparation devra être la règle et le cumul ne pourra pas se réaliser." [1] The same author has also said: "Il s'agit de pouvoirs de droit privé qui tiennent du mandat, et il est préférable de ne pas les confier à un simple pilote, à un chauffeur, comme il a été dit." [2]

Goedhuis speaks of "des pilotes de très bonne qualité, qui par leur ancienneté ont droit à être nommés commandants mais qui ne donnent pas assez de garantie d'instruction générale pour pouvoir réaliser l'importance d'une représentation, telle qu'elle a été prévue dans le projet de Convention." [3]

This objection was expressed on several occasions, particularly in the period prior to World War II, but nowadays it is not heard so frequently, so one may assume that it is no longer valid. In any case it can be refuted by drawing attention to the requirements which present-day airline pilots have to satisfy; these requirements definitely imply an educational standard above the average. Plesman, on the subject "Pilots — then and now," [4] remarks: "Thirty years ago, it was a rather simple matter to become a commercial pilot, because of the relatively short training period required. This, however, presents quite a different picture today. Anyone wishing to become a pilot needs a sound education, which must be followed by at least two years of study at an aeronautical school."

Moreover, in our opinion the powers to be granted to the commander do not call for such a high degree of legal and commercial insight as has been averred. It is rather a question of possessing a technical and practical knowledge of aviation in order to have repairs carried out, to purchase any items that may be required and to make arrangements for the passengers and crew. One may certainly assume that aircraft commanders possess such knowledge.

B. The communication facilities now available make such powers superfluous in the majority of cases.

---

[1] Rapport par M. Babinski (1929), page 7.
[2] Compte Rendu des Réunions de la 4ème Commission, Doc. 84, page 15.
[3] RDILC 1933, page 137.
[4] Plesman, "Thirty years of civil aviation", IATA Bulletin No. 11, 1950 .

Remarks to this effect were made by Ripert [1] in 1928, by de Vos [2] in 1929, by Goedhuis [3] in 1933 and by Niboyet [4] in 1946.

The obvious retort is that in some countries the means of communication are far from perfect, while in any case it takes time and money to obtain advice or authority by cable. In addition, under present world conditions one must allow for the possibility that telegraphic contact may be locally or temporarily interrupted through the outbreak of armed conflicts, local disorders, political upheavals, etc. Writing in 1947, Knauth observed: "It is fair to recall that since 1912 .... there have been World Wars totalling 12 years or one-third of the time, and important wars with cables cut and censorships at inconvenient places, for at least 4 more years, or half the time. It seems, on this record, to be vain to construct a system of operations based on uninterrupted peace and daily and hourly availability of uncensored and uninterrupted communications between the continents." [5]

C. Agents or representatives of the airlines are to be found at practically every stopping place along the air routes.

One of the delegates to CITEJA remarked that: "il trouve choquant le fait que si une compagnie a un représentant à terre, le commandant de l'aéronef puisse par exemple, emprunter de l'argent sans passer par l'entremise de ce représentant". For this reason there was an extremely lengthy exchange of views about limiting the commander's powers in the event of the airline having a local representative.

Apart from the difficulty of formulating such a limitation, it was pointed out that allowance must be made for taxi and charter flights, which often take the aircraft to places where the companies have no local representatives.

Under the system now proposed the operator always retains the right to curtail the powers of the commander in cases where his company has an establishment on the spot.

D. The proposed powers are unnecessarily wide.

The answer to this observation is that the powers in question

[1] Compte Rendu des Réunions de la 4ème Commission (1928), page 18.
[2] Compte Rendu des Réunions de la 4ème Commission, Doc. 2, page 20.
[3] RDILC 1933, page 136.
[4] Doc. 493, page 55.
[5] Knauth, "The Aircraft Commander in International Law" JAL 1947, page 162.

are much less comprehensive than those initially suggested. The power to accept passengers and freight for transportation, to dismiss members of the crew and to represent the operator in law, which appeared in the earlier drafts, have now been cut out.

The latest arrangement gives the commander of an aircraft considerably fewer rights than the master of a ship.

In conclusion it must be emphasized that the objections have largely been met, because, contrary to the original intention, the powers are only granted in order to permit the completion of a trip. As far as repairs are concerned, it is stipulated that these may only take place "to enable the aircraft to proceed promptly on its trip," which indicates that only minor repairs may be carried out. This clause was inserted in response to the comment that the commander seemed to be given the right to have an aircraft repaired even though it was practically a total loss.

An important point on which the draft convention itself is not entirely clear, is whether the operator can limit the powers granted to the aircraft commander. At first the draft included a provision whereby this right was expressly reserved to the operator. Following a long and confused discussion at the meeting of the fourth Committee of CITEJA in Cairo in 1946, however, it was decided to delete this clause, but only after it had been unanimously resolved that the commander's powers might be limited by contract. [1]

In a final peroration the rapporteur outlined his views on the system which he envisaged: A distinction ought to be made in the legal relationship between the aircraft commander and third parties, and between the aircraft commander and the operator. The legal relationship between the commander and third parties is established by the powers specified in Art. 3, which are unchallengeable. Nevertheless the operator can impose certain contractual restrictions on the commander. For example, he may fix a limit to the amounts of money involved, or stipulate that the authority referred to in Art. 3 must not be exercised if a representative of the company is present. The aircraft commander's actions are now binding on the operator at all times, insofar as he has acted within the limits of Art. 3. If the limits laid down in

---

[1] Compte Rendu des Réunions de la 4ème Commission, Doc. 496, page 21.

the contract concluded between the operator and the aircraft commander are exceeded, however, the operator is entitled to recourse against the aircraft commander [1].

After this elucidation of the interpretation favoured by the rapporteur, Art. 3 was adopted by seven votes to two.

The wording of this article is also open to the following comments.

On analyzing what the article really says, ignoring the previous history of it, it may be seen that the commander is granted the right — without special authority — (a) to buy things, (b) to have repairs carried out, (c) to make certain arrangements, (d) to borrow money and (e) to engage personnel.

But these powers are possessed by every private individual; one does not need to be an aircraft commander, nor is it necessary to conclude a convention for this purpose. What is meant of course, although it no longer finds expression in the text, is that the operator shall be bound by such actions of the aircraft commander. This omission is undoubtedly due to the historical development of the present draft. In the original drafts it was made clear that the commander was *the representative* of the operator.

The beginning of this article read as follows: "Le commandant de l'aéronef représente le propriétaire de l'aéronef. Dans cette qualité, même sans mandat spécial, il a le droit: etc. ...." [2] After the War this paragraph again appeared in the draft of the rapporteur Garnault. This met with some opposition in the Committee, however, partly on the following grounds: "On trouve un télégraphe partout. Il ne convient pas de prendre des institutions déjà archaïques dans le droit maritime pour les transporter dans le droit moderne." [3] It was therefore suggested that the first paragraph quoted above should be deleted and that a limitative summary of the powers should suffice.

This was agreed to without much discussion, apparently overlooking the fact that the link between the powers of the commander and the operator had completely disappeared from the text. In our opinion it would be desirable, and more in accordance with the intention of those who drafted the convention, to

---

[1] Compte Rendu des Réunions de la 4ème Commission, Doc. 496, page 44.

[2] Rapport par M. Babinski, Doc. 17, article 8.

[3] Doc. 493, page 55.

amplify the first paragraph in such a way as to show clearly that the commander's actions, within the limits of the specified powers, are binding on the operator.

Otherwise we entirely agree with the contents of this article, though one may wonder whether the power granted under *e* (i.e. the power to engage personnel) will be of much practical use. As each member of the flight crew must have a licence issued in the country where the aircraft is registered, there is very little chance of finding personnel with the requisite qualifications available in foreign countries. In exceptional instances the aircraft commander may perhaps avail himself of this power to engage non-technical personnel (e.g. stewards or attendants to look after animals) as temporary assistants for one flight, but it is questionable whether the latter category of personnel can be regarded as "essential for the trip."

ARTICLE 4

1) The Commander may not, without special authority, sell the aircraft, or, by any contractual act, mortgage or subject it to any similar claim.

Although it had initially been intended to grant the commander very extensive powers as a representative of the operator, the right to sell or mortgage or otherwise encumber the aircraft was expressly withheld [1].

On drafting a limitative list of the commander's powers in private law, it was noticed that there was really no point in maintaining this prohibition. But on the other hand it was pointed out that Art. 3 authorized the aircraft commander to contract loans, so for the sake of clearness it was decided to leave the prohibition in Art. 4 of the draft. The phrase "mortgage or subject it to any similar claim" was employed in order to provide for possible differences in national legislation relating to rights in aircraft [2].

We have no objection to the contents of this article, though in our opinion its inclusion is not strictly necessary [3].

---

[1] Rapport par M. Babinski, Doc. 17, article 9.

[2] Compte Rendu des Réunions de la 4ème Commission, Doc. 493, page 63 and Doc. 496, page 22. For a detailed description of these national regulations see the thesis of Rijks, ,,Het Verdrag van Genève".

[3] This doubt is also expressed by Bucher, op. cit. page 140.

ARTICLE 5.

(1) The beginning and the end of the period during which the Commander maintains disciplinary control over the crew may be fixed by the operator. In any case he is entitled to exercise such control as soon as the crew embarks. At all stopping places, including the end of the trip, he continues to be so entitled at least until the formalities of arrival are completed or until his command is taken over by another person.

(2) The powers of the Commander over the aircraft, the passengers and the cargo on board come into force as soon as the aircraft, with passengers and cargo, are handed over to him at the beginning of the trip. They expire at the end of the trip when the aircraft, the passengers and the cargo have been respectively handed over to the operator's representative or other qualified authority.

This article specifies the period during which the aircraft commander may exercise his authority. Originally the question was settled in very summary fashion by stating that the commander has disciplinary power over the crew "tant qu'il aura besoin de ses services" and over the passengers "tant qu'il se trouve à bord de l'aéronef" [1].

After World War II, at the suggestion of the Greek [2] and American delegates [3] to CITEJA, an attempt was made to define this period more precisely, the aim being to fix the duration of the commander's authority before departure and after landing, and also to clarify the position at the intermediate stopping places. Considerable difficulty was experienced in arriving at a formula which would satisfy practical requirements, as provision had to be made for brief intermediate landings (one hour, for example), overnight stops, and stops where the aircraft was taken over by another crew (the original crew continuing the journey with a different aircraft a few days later). Nowadays the composition of a crew is not the same on all the stages of a long air route flown in accordance with a fixed schedule, as part of the crew may be left behind at an intermediate stop while the rest continue the journey. On the more tiring stretches an extra man may also

---

[1] Compte Rendu de la 6ème Session, Doc. 162, page 171.
[2] Observations par M. E. Georgiades, Doc. 432, page 5.
[3] Compte Rendu des Réunions de la 4ème Commission, Doc. 493, page 65.

be added to the crew (such a crew member flies to and fro on one particular stretch all the time).

Moreover the procedure adopted by the different airlines is not always the same. The station managers or representatives of some companies have far-reaching powers. In certain cases the stewards, for example, may come under the control of a station official. A few of the delegates wanted the commander to retain full authority at all stopping places en route, [1] while others objected that " ... pendant les escales c'est-à-dire que lorsque cet équipage s'en va en ville, le capitaine n'a pas un pouvoir sur l'équipage, étant donné qu'un avion civil n'est pas un avion militaire et n'est pas soumis aux mêmes règles." [2]

Like Art. 2, the ultimate solution makes a distinction between disciplinary control over the crew and power over the passengers, cargo and aircraft. As far as the disciplinary authority over the crew is concerned, the system decided upon allows the operator a certain amount of latitude in fixing the duration of the authority.

Needless to say, this system was not adopted without due consideration; in view of the fact that the disciplinary power is mandatory, it seems strange that the operator should be able to fix its duration. While recognizing the importance of this fundamental objection, we regard the proposed formula as acceptable. Instead of laying down a rigid rule, it gives a flexible arrangement which can be adapted to suit the diverse practical requirements of civil aviation.

Although Art. 5 sets a limit to the period during which the commander's authority may be exercised, to learn the extent of this authority we must refer back to Art. 2, where we find that it applies "within the scope of their duties". In flight, the duties which each member of the crew has to perform are definitely established, and in addition — if necessary — the aircraft commander can assign a crew member to duties other than those which he normally performs. In regard to the duties which each member of the crew is to perform on the ground, however, the various companies have different regulations and at different airports the situation may vary.

---

[1] The Netherlands, Norwegian, Swiss and Egyptian delegates, see Compte Rendu des Réunions de la 4 ème Commission, Doc. 496, page 24 onwards.

[2] Cooper, Doc. 496, page 25.

In some cases, for example, the flight engineers have to carry out repairs on the ground, whereas in other cases this work is done by the ground staff; sometimes the pilots have to prepare a flight plan but it is quite possible that such work may be carried out by the flight operations officer.

We come to the conclusion that the draft convention merely indicates in general terms that the aircraft commander has authority over the crew, and that this vague ruling must necessarily be supplemented by company regulations. The latter regulations ought to contain provisions relating to (a) the duration of the disciplinary control, (b) the duties to be performed by each member of the crew, and (c) the disciplinary measures to be taken by the aircraft commander. If the Convention is ratified, the national legislation will have to make it obligatory for the operator to draw up regulations of this nature.

The power over the aircraft, the passengers and the cargo, begins when they are "handed over" to the commander, and it ends when he in turn has handed them over to the operator's representative or other qualified authority. Here too, the operator is left the necessary freedom of action, as he can specify the moment of transfer.

The transfer can be effected either to a representative or to some other qualified "authority". Since the latter term may denote the local agent or station manager or else the following commander, we feel that the word "authority" should be altered to "person".

Bucher [1] considers these provisions of Art. 5 to be too rigorous as far as the authority over the passengers is concerned; in his opinion it is not correct that the commander should also be able to exercise authority over the passengers during the intermediate landings.

We disagree with Bucher for the following two reasons. Firstly, because if the intermediate landing is to last some time, as in the case of an overnight stop, in practice the aircraft commander will hand over the aircraft, with passengers and cargo, to a qualified member of the ground personnel immediately after landing, thereby relinquishing his powers at the same time. Secondly, because on the ground the commander's authority over the passen-

---

[1] Bucher, op. cit. page 141.

gers is only nominal if such a transfer does not occur in some exceptional case; as the power is exclusively intended to ensure order and safety, the commander cannot use it except in very special contingencies.

ARTICLE 6

(1) In all countries and under all circumstances the Commander shall have the right of access to:
a) the Consul of the nationality of any person on board;
b) the Consul of the State in which the aircraft is registered;
c) the Consuls of the States of either party to a charter of the aircraft, and of consignors of cargo.

(2) After hearing the Commander, the Consuls may take any necessary means which are in accordance with the laws and consular regulations of their respective States.

(3) If the Commander first refers to the Consul of a State other than the State in which the aircraft is registered, the Commander shall notify this action as soon as possible to the nearest Consul of the State in which the aircraft is registered.

These provisions were inserted in the draft in 1946 at the instance of the United States delegates. The discussions in CITEJA were mostly concerned with the form of the proposed rules. There was little difference of opinion as to the principle. In their writings on the subject, however, some authors have questioned the usefulness of including such regulations. [1] From the minutes of the discussion, it appears that the United States delegate Knauth was thinking of the case where an aircraft commander is detained by the local police, e.g. after an accident. Knauth wished to make sure that in such a situation the aircraft commander would always have direct access to a consul who could then take whatever steps might be necessary.

Knauth, quite correctly, did not want the steps in question to be specified in this Convention: "Nous ne désirons donc pas nous immiscer plus avant quant aux questions consulaires, mais nous voulons nous limiter à la question du droit de passage par

---

[1] Honig, ,,De positie van de gezagvoerder van een luchtvaartuig", NJB 1951, page 317; Bucher, ,,Le Statut juridique du Personnel Navigant" page 141; Riese, ,,Luftrecht" page 220.

intermédiaire du gendarme au consul et que le consul veuille bien ouvrir la porte." [1] The American proposal was adopted by the rapporteur Garnault, and subsequent versions of the draft contain such phrases as "faire appel à la protection ...." [2] and "solliciter la protection ...," [3] while the draft finally agreed upon by CITEJA speaks of "droit d'accès auprès ...." [4]

The latter wording, which also appears in the present ICAO draft, best expresses the original idea of granting the commander right of access to a consul, in a physical sense, under all circumstances.

Which consuls can the aircraft commander approach? Not only the consul of his own nationality but also the consuls enumerated in Par. 1 (a ,b and c). It will be noticed that the consul of the State of the operator is not mentioned. In our opinion the argument put forward [5], that the operator is usually not present in person is not at all convincing, as the same thing may apply to the parties to a charter agreement and the consignors of cargo. Par. 1 (c) should therefore be amended to include the consul of the operator as well.

The rule laid down in Par. 1 does not appear to be absolutely necessary, but it may indeed be useful in certain cases. Accordingly, we should like to retain this first paragraph in the draft. The second paragraph, however, is superfluous; it is obvious that the consuls may take the measures which are in conformity with the applicable laws and regulations. Such a provision in the present Convention therefore serves no useful purpose. Lastly, Par. 3 gives a detailed regulation which is equally out of place in this Convention, and in our opinion it too could easily be deleted.

ARTICLE 7

1) Births and deaths occurring on board the aircraft shall be recorded in the journey log-book by the Commander, who shall issue extracts to the parties interested. He shall as soon as possible transmit certified extracts to the competent authority of the State in which the aircraft is registered and to that of the place of first landing, if so requested by the local authorities.

---

[1] Compte Rendu des Réunions de la 4ème Commission, Doc. 496, page 31.
[2] Rapport et Avant-projet par M. Garnault, Doc. 451, art. 21.
[3] Avant-projet de Convention, Doc. 471/C article 6.
[4] Compte Rendu de la 15ème Session, Doc. 485 page 95, article 6.
[5] Doc. 496, page 32.

In contrast to the preceding article, the subject matter of this article has long been a point of discussion. Maritime law furnished a large measure of inspiration in the early stages, so it was only natural that the conclusion should soon be reached that the commander of an aircraft ought to have the same powers as the master of a ship to act as a sort of registrar.

At a Congress held as far back as 1912 it was decided that in the event of a birth or death on board an aircraft, the commander must record the occurrence in the log-books and give a copy of the entry to the authorities at the first place of landing. [1] This question was not discussed in the first report submitted by Thieffry, which was based entirely on private law, but Babinski sent out a questionnaire asking for the opinion of the delegates on this point. It was agreed in principle that the aircraft commander ought to be given some powers in this respect, but there was considerable difference of opinion as to the extent of these powers. The following occurrences in connection with which the aircraft commander might have to fulfill some role were mentioned: birth, death, marriage and the making of a will. Generally speaking, the idea of granting power in the last two cases was turned down. With regard to birth and death, there was a cleavage of opinion between those who wanted the aircraft commander to draw up a simple report of the occurrence, and those who wished to give him the power to make out an official certificate. [2]

In his preliminary draft Babinski proposed that powers should only be granted for cases of birth and death, and that it should be left to the States to decide the extent to which they would grant the aircraft commander powers as a registrar. [3] An optional system of this nature may lead to all sorts of complications and it therefore aroused a lot of criticism. [4]

In the rapporteur's next draft the comments were taken into account and a mandatory system was prescribed. Instead of being given the status of a registrar the aircraft commander merely got the power (and duty) of recording the

---

[1] Deuxième Congrès du Comité Juridique International de l'Aviation, Genève 1912, page 155.

[2] Compte Rendu des Réunions de la 4ème Commission, Doc. 2, page 21.

[3] Rapport par M. Babinski, Doc. 17, page 9.

[4] Compte Rendu des Réunions de la 4ème Commission, Doc. 37, page 45 onwards and page 57 onwards.

occurence in the log-books and issuing extracts to the parties interested. [1] This text was adopted by the 4th Commission subject to the addition of a clause stipulating that extracts must be issued to the competent authorities of the place of landing and of the State of registry of the aircraft. [2]

A similar text was agreed upon by CITEJA and it appeared in the draft which was provisionally adopted in 1931. [3]

When study of the draft was resumed in 1946, the principle established in 1931 proved to be still acceptable in every way. The only new point was that the United States delegate wanted the text to state explicitly that the aircraft commander was forbidden to perform marriages or to act as a notary. [4]

In order to meet this wish, a clause to that effect was included in the text. Objections were subsequently raised, however, as some delegates were anxious to leave the way open for the commander to perform a marriage *in extremis* [5] while others did not want to rule out the possibility of drawing up a legally valid will under special circumstances [6]. Moreover, an express prohibition seemed to be superfluous, since the first paragraph of this article specifies what the commander is entitled to do. The clause relating to marriages and wills was therefore deleted again. [7]

We consider that the contents of this article are perfectly acceptable. The commander's role was rightly confined to birth and death, events "qui se réalisent indépendamment de la volonté de l'homme" [8].

In our opinion it is also correct that the aircraft commander should merely have to establish and record the events referred to. It is left to the competent authorities on the ground to draw up a legally recognized document. An additional benefit of this procedure is that the extremely controversial question of the

---

[1] Rapport Supplémentaire par M. Babinski, Doc. 67, page 5.
[2] Compte Rendu des Réunions de la 4ème Commission, Doc. 84, page 45.
[3] Compte Rendu de la 6ème Session, Doc. 162, page 95.
[4] Compte Rendu des Réunions, Doc. 493, page 75.
[5] The Belgian delegate de Smet, Doc. 496, page 36.
[6] The Egyptian delegate Francis, Doc. 496, page 35.
[7] ibidem, page 35.
[8] Babinski in Doc. 17, page 9.

nationality of a child born on board an aircraft [1] is kept out of the Convention.

The proposed regulation is likewise regarded as being generally acceptable in the writings on this subject [2].

ARTICLE 8

1) The provisions of this Convention do not affect any international conventions or the laws or regulations of the Contracting States defining the conditions of qualifications required of an aircraft Commander.

This article was inserted because some of the CITEJA delegates felt that the draft Convention tended to make the aircraft commander a man of importance, vested with rights and obligations, and that he should therefore be required to satisfy certain minimum standards — especially as the choice of the aircraft commander was left to the operator. It was desired to emphasize that the operator was not absolutely free in his choice, since the aircraft commander designated by him must meet the relevant requirements.

This Convention was not considered to be the appropriate place for regulations concerning the commander's qualifications. It was therefore decided that the regulations in question should be drawn up by CINA or ICAO or the national legislators, while the article itself would merely refer to these regulations. An article of this tenor — referring only to the national law — was included in the first draft of Babinski [3] and it is now repeated in essentially the same form in Art. 8 of the ICAO draft which is quoted above.

ICAO has promulgated a large number of technical regulations in the Annexes to the Chicago Convention [4], but the term "aircraft commander" is not used in any of these Annexes. Instead they speak of the "pilot-in-command." We must assume that the latter official should be assimilated with the "aircraft commander" referred to in the draft Convention.

It is difficult to arrive at any other conclusion, since it is stated that the aircraft commander "shall be in charge of the

---

[1] See Lemoine, "Traité de Droit Aérien", page 204; Coquoz, "Le Droit Privé Aérien," page 292.

[2] Bucher, op. cit. page 142; Honig, NJB 1951, page 317.

[3] Rapport par M. Babinski, Doc. 17 art. 20.

[4] See page 12.

aircraft, the crew, the passengers and the cargo," whereas the pilot-in-command "shall be responsible for the operation and safety of the aircraft and for the safety of all persons on board, during flight time." [1] If these definitions refer to two different persons, it would mean having "two captains on one ship." We are therefore obliged to conclude that the aircraft commander and the pilot-in-command are one and the same person. Thus the operator is not so free in his choice of the commander as Art. 1 of the Convention would suggest. It follows from the foregoing that he must designate as commander of the aircraft a pilot who meets the requirements set out in the applicable regulations. [2]

In our opinion it is advisable to bring the terminology of the draft Convention into line with the terminology of the existing rules and regulations of ICAO.

The term "aircraft commander" is preferable in this connection particularly in view of the regulations governing the order of succession (Art. 1, Par. 3). One can well imagine that under exceptional circumstances a flight engineer or radio operator might take over the duties of the aircraft commander, but it is most unlikely that he could act as pilot-in-command. Moreover, pilotless aircraft will not have any pilot-in-command on board; as we have already remarked [3], however, it appears to be desirable that in certain cases (when passengers are carried) there ought to be an aircraft commander on board. Should the expression "pilot-in-command" be replaced by "aircraft commander" in the Annexes, it will of course be necessary to indicate in some other way that the pilot acts as commander of the aircraft.

For the rest we have two comments regarding the train of thought disclosed in Art. 8.

In the first place, this article refers to "international conventions or the laws or regulations of the Contracting States." In our opinion a reference to the national legislation in a convention of this nature is generally undesirable because it tends to stand in the way of the vitally necessary unification of air law. Furthermore, in this particular case such a reference seems superfluous, as the same result can be achieved by a mere reference to "inter-

---

[1] Annex 6, par 4.5.1.

[2] The regulations recommended by ICAO are given in detail in Annex 2 and Annex 6.

[3] See page 133.

national conventions." After all, Art. 32 of the Chicago Convention states that: "The pilot of every aircraft and the other members of the operating crew of every aircraft engaged in international navigation shall be provided with certificates of competency and licences issued or rendered valid by the State in which the aircraft is registered."

It is obvious that this conclusively establishes the power of the national authorities to apply their own standards for the qualifications of an aircraft commander.

In the second place, one may wonder why reference is only made to other regulations in connection with the *qualifications* of the commander. Other rules which affect the status of the aircraft commander — especially some constituting obligations — are to be found in existing international conventions or in the regulations issued pursuant to those conventions.

Since Art. 8 expressly states that the laws or regulations governing the commander's qualifications are *not* affected, it might well be reasoned by contraries that the obligations defined elsewhere *are* affected by this convention. In other words, they may become null and void. Such an assumption might lead to very unacceptable consequences, as may be seen from the following examples.

1. We have already explained [1] that in general the aircraft commander is obliged to comply with the instructions given by A.T.C. in order to avoid the risk of collision. In our opinion this obligation is difficult to reconcile with the clause in the draft convention which states categorically that the commander shall be in charge of the aircraft. It seems extremely undesirable that the aircraft commander should be able to disregard the authority of A.T.C. [2] in virtue of the power conferred on him by this convention.

2. According to the draft, the aircraft commander is authorized to disembark any member of the crew or the passengers "for good reason." Suppose that one of the passengers proves to be suffering from an infectious disease. This is undoubtedly a "good reason" for the aircraft commander to get rid of him by putting

---

[1] See page 54.

[2] A remark to this effect was also made in the U.K. commentary on the draft in C-WP/899, page 15.

him off the aircraft at the first opportunity. Is the aircraft commander now entitled to ignore the provisions of the International Sanitary Convention which stipulate that in such a case — subject to the consent of the local authorities — disembarkation may only take place at an aerodrome equipped with quarantine facilities? [1] To our mind this would be utterly out of the question.

3. According to Art. 3 (e) of the draft, the commander may engage "such personnel as is essential for the completion of the trip" in order to replace members of the crew who are no longer available. There is nothing to indicate that such personnel must be in possession of certificates of competency for the duties to be performed. Since Art. 8 expressly stipulates that the aircraft commander must possess qualifications specified elsewhere, one might deduce that such a rule apparently does not apply to the personnel engaged by him. This would be extremely undesirable with a view to safety in air navigation.

There are two possible ways of avoiding the above difficulties:

a. Art. 8 might be amended to read as follows: "The provisions of this Convention do not affect any international Conventions pertaining to the qualifications, rights or duties of the aircraft commander or to the operation or use of aircraft".

b. An alternative solution would be to reason that this convention is intended to amplify the law and not to conflict with other conventions, in which case Art. 8 would have to be completely omitted in order to obviate any misunderstandings.

We find the latter solution preferable. Anyhow, in view of the foregoing, it does not appear right to retain the article in its present form.

ARTICLE 9

(1) This Convention shall be applied in the case of an aircraft performing an international flight provided that such aircraft is registered in one of the Contracting States or is operated by a national of any such State.

---

[1] cf. arts. 24 and 25 of that Convention.

ARTICLE 10

(1) This Convention does not apply to the Commander serving on board military, customs or police aircraft.

These articles jointly indicate the cases in which the Convention is applicable. Summarizing the two articles, it appears that three conditions must be satisfied if the Convention is to apply.

1. It must be an international flight;
2. The aircraft must be registered in one of the Contracting States, or else the operator must be a national of such a State;
3. The aircraft must be a civil aircraft.

When discussing Art. 1 we drew attention to the absence of a definition for the term "international flight," and at the same time we stressed the desirability of making the Convention applicable to non-international flights as well as international flights. [1]

We have no comments regarding the second condition except to say that the wording should perhaps be altered to cover the frequent cases where the operator is not a natural person but a corporate body.

The third condition brings us back to the use of the term "State aircraft" as opposed to "civil aircraft." [2] The description used is the same as in the Chicago Convention. In our opinion the wording could be improved by omitting the words "the Commander serving on board" from Art. 10.

ARTICLES 11–19

Arts. 11–19 contain formal provisions relating to the ratification, effectiveness and denunciation of the Convention.

As they are not of immediate importance for the legal status of the aircraft commander, we shall not discuss them in detail.

The full text of the draft convention is appended to this study in the form of an Appendix [3].

---

[1] See page 132.
[2] See page 24.
[3] See page 171.

CHAPTER III

ADDITIONS TO THE DRAFT

On analyzing the draft Convention in the previous chapter, we found that on the whole it was acceptable — apart from a few minor additions and amendments.

However, as the draft is intended to regularize the complete legal position of the aircraft commander, we must point out that it does not cover several important aspects of his legal status. A convention of this nature is bound to take the form of a compromise. Certain controversial points were completely omitted owing to the difficulty of reaching agreement. In judging the desirability of possible additions to the draft it must therefore be borne in mind that such additions may perhaps prevent universal acceptance of the Convention.

LIABILITY OF THE AIRCRAFT COMMANDER

Regulations concerning liability were initially — and rightly — regarded as being part and parcel of the legal status of the aircraft commander. Accordingly, numerous attempts were made to lay down rules on the subject. [1]

Thieffry's draft, which appeared in 1927, contained the following provision: " ... Toutefois le capitaine est garant de ses fautes même légères dans l'exercice de son mandat." [2]

A difference of opinion on this point at once became evident during the first discussions. The British delegate, Sir Alfred Dennis, considered that in view of the position of aeronautical development it was unreasonable to impose such a heavy liability on the aircraft commander, especially as other Commissions of

---

[1] The statement made by Honig in NJB 1951, page 317, that in the drafts published up to now no rules are laid down regarding liability, is untrue.

[2] Avant-projet de Convention par M. Thieffry (1927), art. 2.

CITEJA were drafting a limitation of the liability of the operator and carrier [1].

The French delegate Ripert also considered it unreasonable that the aircraft commander's liability should not be limited in any way. He wondered how anybody could be willing to act as commander under such a burden of personal liability, remarking:

"Ce qui paraît très grave, c'est de dire qu'un commandant d'aéronef, pour le salaire qu'il recevra, prendra la responsabilité personnelle de tous les voyageurs, de toutes les marchandises, pour la moindre faute, sans que sa faute soit absorbée par les risques de la navigation qui jouent en faveur du transporteur. Il n'a pas les bénéfices de l'exploitation, et il prend toute la charge de la responsabilité!" [2]

In view of these criticisms the rapporteur Babinski inserted the following clause in his draft: "En ce qui concerne la responsabilité du commandant de l'aéronef envers les passagers, les chargeurs et, en général toute tierce personne, il n'est tenu personnellement qu'au cas de faute volontaire délictuelle; s'il s'agit d'une faute de fonction, c'est la responsabilité du propriétaire et non la sienne qui se trouve engagée". [3]

The Swiss delegate Pittard agreed in principle that the liability of the commander should be limited, but he proposed a different wording which only made the commander liable if there was any question of "un acte illicite intentionnel."

The Swedish delegate Engstromer, however, was very strongly opposed to this idea; he remarked: "C'est presque le dol çela; ce n'est pas assez et il est dangereux de limiter la responsabilité du commandant au cas de dol. C'est dangereux parce que le commandant doit être très prudent. Il doit avoir un règlement qui le rende aussi prudent que possible. Si on borne sa responsabilité au dol, on aboutira au résultat contraire. Il se dira: je ne suis pas responsable, je puis faire ce que je veux!" [4]

The counter-argument was that it is in the general interest that

---

[1] The relative proposals have in the meantime been incorporated in the Conventions of Warsaw and Rome, see pages 88 and 94.

[2] Compte Rendu des Réunions de la 4ème Commission, Doc. 2, page 11 onwards.

[3] Rapport par M. Babinski, Doc. 17 art. 17.

[4] Compte Rendu des Réunions de la 4ème Commission, Doc. 37, page 49 onwards.

airlines should be liable for the actions of the commanders of their aircraft, as this will oblige them to appoint experienced and careful commanders; it is better for the public that a heavy responsibility should be imposed on the carriers rather than on the commander, since the latter will usually be financially incapable of paying the large sums which may be involved in accident claims.

Whatever the rights and wrongs of the matter, Babinski changed his tack completely and the next draft included an article reading as follows:

"L'exploitant aura toujours un droit de recours contre le commandant de l'aéronef en prouvant son dol ou sa faute, si, en vertu de textes en vigueur, il a été condamné à payer, voire assigné en paiement d'une somme pour dommages causés aux tiers du fait du commandant." [1]

Naturally this article proved to be unacceptable to the British delegate. He said: "Si on prend la question de fond: un commandant ou un pilote commet une faute de navigation grave: il va payer une somme de x à son patron pour sa faute. Mais quel est le pilote qui consentira à entreprendre un voyage dans de pareilles conditions? Quel pilote acceptera de s'exposer à payer des sommes peut-être considérables pour une faute? C'est pratiquement impossible. Les exploitants eux-mêmes semblent accepter les fautes de navigation dans certains cas; dans la navigation maritime, les commandants sont généralement exemptés de fautes de navigation."

He suggested that the article should be deleted and that the question of liability should be settled by the national legislators [2]. The Commission had got into an impasse; at the 6th meeting of CITEJA, which was held in 1931, the article on liability was rejected by 13 votes to 1, with 1 abstention [3].

There are three important points to be taken into consideration in judging whether rules of liability ought to be added to the Convention.

1. Is this Convention an appropriate place for such rules?

Although the Russian delegate to CITEJA expressed the contrary

---

[1] Rapport Supplementaire par M. Babinski, Doc. 67, art. 12.
[2] Compte Rendu des Réunions de la 4ème Commission, Doc. 84, page 40.
[3] Compte Rendu de la 6ème Session, Doc. 162, page 92.

view, in our opinion this question must be answered in the affirmative. After all, liability forms a very important aspect of the legal status of the aircraft commander, and the value of the Convention is therefore seriously reduced through the absence of rules of liability.

2. Is there much chance of international acceptance of such rules?

Originally it was argued that common law should apply to liability in aviation, but this view has now been superseded. Special rules governing the liability for the carrier and operator have already been agreed to in the Warsaw Convention of 1929 and the Rome Convention of 1933. As far as can be deduced from the minutes of CITEJA, there was considerable support for the standpoint that the liability of the aircraft commander should similarly be limited, but the delegates disagreed as to the wording of such a limitation. A system which would have been very onerous for the aircraft commander was rejected by a large majority.

As the problem has not been dealt with in the framework of this project during the past 20 years, it is difficult to predict how it will be viewed nowadays. The standards of knowledge and skill which aircraft commanders are required to satisfy have been raised considerably since 1931, but the number of aircraft accidents ascribed to pilot error has by no means diminished, [1] so the problem is definitely one of current importance.

The question of the commander's liability has also arisen in connection with the revision of the Warsaw and Rome Conventions by the ICAO Legal Committee. [2]

Considering also that the members of IATA expressly exclude the liability of employees in their conditions of carriage and that IFALPA is pressing for settlement of the matter, in our opinion it may be assumed that there is a fairly good chance of international acceptance of a set of rules governing the liability of the aircraft commander.

3. How should such rules be formulated?

In 1950 IFALPA adopted a resolution reading as follows:

"It is resolved that a captain of aircraft shall not, in principle, be held liable for loss or damage to property, cargo and valuables

---

[1] See page 98.
[2] See pages 92 and 96.

11

carried in aircraft under his command either by passengers or third parties resulting from accident or otherwise either in the air or on the ground and that in so far as he might be held liable this liability shall not exceed that of the operating company and shall be fully covered by an insurance policy taken out by the operating company."

The arguments put forward were that: "In the first place, if it is proper that the operator should be safeguarded against the possibility of financial ruin as a result of action arising out of an accident, it is also proper that an employee and his estate should also be protected", and "Secondly, excluding the pilot from liability for accident will, by contributing to his peace of mind, actually tend to prevent the occurrence of accidents." [1]

The latter argument appears to have received support from other quarters. It has been remarked that "subconscious distraction created by future economic insecurity is a flight 'hazard' from which the pilot should be relieved." [2]

It seems that the wish of IFALPA has already been met by some airlines because they have arranged for their insurance policies to cover the personal liability of the aircraft commander.

In view of the fact that a similar clause or interpretation may not be in general use as yet, we should prefer to have the liability of the aircraft commander limited by statute as a matter of principle. We think that the Convention ought to include a provision similar to Art. 342 (2) of the Dutch Commercial Code, whereby the master of a ship is only held liable for damage which he causes to others through *wilful misconduct or gross negligence* in the course of his employment.

In the explanatory memorandum appended to the bill, this limitation of liability was justified on the following grounds:

"The consideration which led to the insertion of this provision is that the mistakes made by the Ship's captain often consist of errors in judgment under difficult and perilous circumstances. Every captain, even the most experienced, is liable to make such errors, because human judgment is not infallible. It would be inhuman and unjust to expose the captain, who already exercises a dangerous and responsible profession, to the risk of losing his

---

[1] ICAO Doc. A4 — WP/154.
[2] Flight Safety Foundation, Accident Prevention Bulletin, 50—8.

savings and being reduced to beggary if it is established that he committed an error of judgment in a moment of confusion when disaster was imminent." [1]

This is indeed a reasoning which, *mutatis mutandis*, may also apply to the commander of an aircraft.

Honig [2] has suggested the following text: "The commander acting within the scope of his authority is liable for damage caused by him to others through a deliberate act or omission." In our opinion this is acceptable, subject to insertion of the word "only" after "liable".

DUTIES OF THE AIRCRAFT COMMANDER

The draft convention grants the aircraft commander a number of rights but it says practically nothing about his duties. As with the rules of liability, the draft originally contained a clause relating to the commander's obligations, but that clause was later deleted.

The following article appeared in the draft provisionally adopted by CITEJA in 1931:

"Le Commandant de l'aéronef doit veiller à la sécurité du voyage; l'aéronef reste pendant le voyage sous sa surveillance et sa responsabilité, et le Commandant ne peut pas le quitter de sa propre volonté sans motifs graves."

As the rapporteur pointed out, safe conduct of the flight ought to be one of the principal obligations of the aircraft commander, and the powers granted to him really arise out of this obligation. [3]

The wording of the second part of the article is not so clear and it has therefore been criticised: If the aircraft commander wants to have a meal in the airport restaurant or to spend the night in a hotel after completing one stage of a journey by air, can this be regarded as a "motif grave" for leaving the aircraft? [4]

The original wording was almost entirely retained when this article was included in Garnault's draft in 1946.

Two distinct lines of thought became apparent during the subsequent discussions.

---

[1] Cf. Cleveringa, ,,Het Nieuwe Zeerecht", page 233.
[2] NJB 1951, page 317.
[3] Avant-Projet de Convention par M. Babinski, Doc. 119, page 12.
[4] Goedhuis, RDILC 1933, page 142.

On the one hand it was held that the duties of the aircraft commander were not specified in sufficient detail, [1] while on the other hand it was commented that these duties had already been elaborated in other regulations, notably the Annexes to the Chicago Convention. [2] In the end it was decided to scrap the whole article.

This decision is regrettable. Certainly it is better to give detailed regulations concerning the technical and administrative duties of the aircraft commander by means of the Annexes, which do not constitute mandatory law and can easily be altered, but we consider that a duty of fundamental importance, viz. the duty to ensure that the flight is conducted with safety, must not be omitted from a convention on the status of the aircraft commander.

We feel it advisable that the Convention should include a general provision to the effect that the commander is under an obligation to perform his task properly and conscientiously.

Such an obligation ought to attach to the function of aircraft commander as inseparably as the rights granted in this Convention.

In accordance with the rule laid down for the master of a ship in Art. 342 of the Netherlands Commercial Code, it has been proposed that an article reading as follows should be included in the Convention: "The commander is responsible for the operation of the aircraft. He is obliged to act with the skill, exactness and discretion necessary for a proper fulfilment of his task." [3]

We are in favour of adding an article of this tenor to the Convention.

POLICE POWERS

Various writers are in favour of granting police powers to the aircraft commander in case an offence or crime is committed on board. [4]

Swiss air law contains a provision of this nature, the commander being obliged "die Spuren der Tat fest zu stellen und zu sichern, dringliche Untersuchungshandlungen vorzunehmen und nötigen-

---

[1] Observations par M. Georgiades, Doc. 432, page 5.

[2] To this effect Garnault in Doc. 434, page 6; Wilberforce in Doc. 493, page 73; de Smet in Doc. 496, page 30.

[3] Honig, loc. cit.

[4] Honig, loc. cit.; Maschino, op. cit. page 116; Riese, "Luftrecht", page 220; Wüstendorfer, ArchfLR 1931, page 212; Bucher, op. cit. page 149.

falls einen Verdächtigen vorläufig festzunehmen, als Beweismittel
geeignete Gegenstände zu beschlagnahmen und die Flugzeugin-
sassen zu durchsuchen."[1]

To our mind, the inclusion of such far reaching powers, which
are strongly reminiscent of the regulations in force in maritime
law,[2] is most inexpedient. How can anybody be taken into custody
on board an aircraft, and why can the searching of the persons on
board not be postponed until after the landing?

The brief duration of a journey by air and the primary duty of
the aircraft commander to ensure the safe conduct of the flight,
will often prevent him from instituting a formal investigation on
board the aircraft. As the persons on board — and the corpus
delicti, if any — can hardly disappear or be removed during the
flight, especially with modern aircraft, in our opinion there is not
the slightest objection to having the investigation carried out
by the competent police authorities after the landing.[3]

All that appears to be necessary, namely the power to maintain
order and safety, is already granted in Art. 2 of the draft Convention. If so desired, one might make it obligatory for the aircraft
commander (a) to land as soon as possible after finding that a
crime or offence has been committed, and (b) to report the
occurrence to the appropriate authorities by radio if possible but
in any case immediately after the landing. Such a regulation
would be roughly the same as the regulation concerning notification of birth and death on board an aircraft.

The more comprehensive proposal that the aircraft commander
should be given the status of a police or criminal investigation
official would undoubtedly encounter objections similar to those
raised when it was suggested that the aircraft commander be
made a sort of registrar[4]. Apart from the fact that, as already
remarked, such far-reaching power is of little practical value, it
would first be necessary to find an internationally acceptable
solution for the problem as to which jurisdiction applies on board
aircraft. ICAO is still studying the problem of the legal status

---

[1] Riese, op. cit. page 221.
[2] cf. Cleveringa, ,,Het Nieuwe Zeerecht" page 254.
[3] The case given by Knauth (ILA Legal Committee, Lucerne Conference 1952, page
25) of a criminal escaping by parachute is of a hypothetic nature.
[4] See page 151.

of aircraft [1]. For the above reasons we consider it inadvisable to include any regulations with respect to the police powers of the aircraft commander in the Convention; at the most, although it is really unnecessary, one might impose a simple obligation along the lines indicated above.

---

[1] Cf. Legal Committee, Minutes and Documents 8th Session, Doc. 7229—LC/133.

SUMMARY

Air law, to a greater degree than most other branches of law, is
not yet stabilized. As the history of this new branch of jurisprudence only covers a few decades and the future development is
difficult to foresee, it is perhaps inevitable that a study in this
field should bear the character of a snapshot. The object selected
for this purpose in the first part of the present study was the
aircraft commander under existing Dutch law.

In public law the aircraft commander is subject to very
extensive rules and regulations whereby countless duties are
imposed on him in the preparation and execution of a flight as
well as after its completion. Since the aircraft commander can
enter and leave several jurisdictions within a short space of time,
unification of these rules and regulations is very important. The
"Standards and Recommended Practices" established by ICAO
in the Annexes to the Chicago Convention are not mandatory,
however, and although they are generally implemented — subject
to a few deviations — they often lack the force of law. Dutch air
legislation, which only applies above Dutch territory, is in an
inconvenient stage of transition, as it has not yet been adapted
to the ICAO standards.

In some cases, e.g. the keeping of flight documents and compliance with public health regulations, the aircraft commander
has to perform certain duties as an agent of the authorities. In
this respect, however, his duties are considerably fewer than those
of the master of a ship, since the latter has police powers and has
to act as a registrar of births and deaths, etc.

As far as the lacunae in the legal status of the aircraft commander are concerned, attention was drawn to the absence of a
general statutory power to exercise authority over the persons
on board. The aircraft commander's power to act as a representative of the operator was also found to be without a legal
foundation.

On studying the vehicle in which the aircraft commander performs his daily work, i.e. the transport plane, we found how difficult it is to make a clear distinction between civil and military aircraft. It was also found that aircraft possess a nationality but that there is considerable cleavage of opinion regarding the legal jurisdiction on board the aircraft.

According to current Dutch law, one of the members of the crew is the commander of the aircraft, but this need not be a pilot. It was pointed out that licences are issued to applicants who satisfy the requirements in respect of physical fitness, knowledge, and technical skill, whereas qualities of character are ignored. We came to the conclusion that the Dutch legislation on this point ought to be amended.

The independence of the aircraft commander has been greatly reduced through the development of telecommunication facilities which permit constant contact with services on the ground throughout the flight. Attention was devoted to the power of Air Traffic Control, Operational Control and the airport manager to issue instructions to the aircraft commander during the flight. The question of the aircraft commander being obliged to comply with instructions from the ground is particularly important when he has to decide whether to attempt a landing under unfavourable weather conditions.

The commander's duties in connection with search and rescue are described in the Brussels Convention (which has not been ratified) and in Annex 12 to the Chicago Convention. The drawbacks and advantages of the two sets of regulations were analyzed and we came to the conclusion that it is desirable to combine their most useful elements.

Sanctions can be imposed on the aircraft commander by criminal courts in the Netherlands and abroad while disciplinary measures can be taken by the Dutch Air Accident Board. The latter body can also withdraw a crew member's licence if it considers that the person in question is not competent to perform his duties. Although the composition and the procedure of the Air Accident Board provide a number of guarantees for correct administration of justice, at present there is no right of appeal to a higher court. This is a shortcoming which ought to be remedied.

The liability of the carrier and the operator is governed by the

Warsaw Convention and the Rome Convention, but the liability
of the aircraft commander is not (or only partly) settled by these
Conventions. As a result of this the aircraft commander's liability
is frequently determined by the general rules of law. In our
opinion the special character of aviation makes it desirable to
have separate comprehensive regulations on the subject. On
account of the relatively high frequency with which aircraft
accidents are ascribed to pilot error, we considered the standards
to be applied in this connection and also the effect on the liability
of the carrier, operator and aircraft commander.

It was found that the aircraft commander's status as an employee
was also governed by common law. Although the circumstances under which the commander performs his duties have
a number of exceptional features, we do not think there is any
necessity for special statutory regulations on the subject.

In our opinion it may be concluded from the foregoing that
the legal status of the aircraft commander has many gaps and
uncertainties, and that his position is often "in the air" both
literally and figuratively.

If one regards an aircraft in flight — together with all the
persons and goods on board — as a small community temporarily
cut off from the rest of society, it is clear that this community
needs to have some rule of law. In view of the international
character of civil aviation, it is desirable that such a rule of law
should apply all over the world.

The second part of our study was devoted to the attempts to
prepare a convention on the legal status of the aircraft commander.
For more than twenty-five years this subject has been under
consideration by CITEJA and now by ICAO, and numerous
drafts have appeared during this period.

Owing to the lack of precedents in air law, the status of the
aircraft commander has been compared with that of the master of
a seagoing vessel, not to mention that of a captain of a river
vessel, a bus driver or a railway guard.

All of the vehicles involved are capable of traversing international
frontiers and covering long distances with a comparatively
isolated group of people on board.

It appears to us, however, that there are more points of difference than of resemblance in these comparisons.

In the case of inland shipping, buses and trains, the isolation of the persons on board is considerably less than on an aircraft; the conveyances just mentioned are always on the territory of other states, and never on stateless territory; they move along fixed routes; the State concerned can always exercise real authority and intervene in the jurisdiction on board at all times.

On comparing the aircraft commander with the master of a ship, however, a certain analogy is unmistakable. And yet there are differences of principle and degree: the duration of a journey by air is generally much shorter; the crew of an aircraft is considerably smaller than the crew of a seagoing vessel, and the number of passengers usually less; the limited freedom of movement on board an aircraft gives the community a fundamentally different character from that on board a ship; tradition, which plays such an important part in shipping, has not yet been established in aviation.

For all these reasons it seems advisable that a statute for the commander of an aircraft should not be based on seeming resemblances between air transport and other means of transport; on the contrary, the aircraft ought to be looked upon as a vehicle *sui generis* and the commander given a legal status appropriate to the special features of civil aviation.

The draft convention of ICAO on the legal status of the aircraft commander appears to satisfy this requirement and on the whole it is acceptable. But in our opinion the Convention should also be applicable to non-international flights.

Apart from a number of additions and amendments of minor importance, we think the draft should be amplified in two respects. In the first place we feel that the Convention ought to contain a clause establishing the commander's duty to ensure that the flight is conducted with safety. Secondly, we consider that the liability of the aircraft commander ought to be confined exclusively to cases where there is some question of wilful misconduct or gross negligence.

We end our observations by expressing the wish that a Convention may soon be concluded to grant the aircraft commander a legal status in keeping with the important and responsible task which he performs in commercial air transportation.

INTERNATIONAL CIVIL AVIATION ORGANIZATION

DRAFT CONVENTION ON THE LEGAL
STATUS OF THE AIRCRAFT COMMANDER [1]
(as revised by the Paris Legal ad hoc Committee)
(February 1947)

THE UNDERMENTIONED GOVERNMENTS,

CONSIDERING it desirable that the status of the Commander of aircraft in international flight should be defined and regularized, and
HAVING AGREED upon certain principles and arrangements in connection with this matter,

HAVE ACCORDINGLY DECIDED TO CONCLUDE THIS COVENTION:

Article 1.

(1) Every aircraft performing an international flight shall carry one person vested with the powers of a Commander.
(2) The right to designate the Commander belongs to the operator of the aircraft.
(3) In the absence of any Commander so designated, or in case the latter is prevented from performing his duties, and if no successor has been designated by the operator, the Commander's duties will be carried out by the other members of the crew in the following order: pilots, navigators, engineers, radio operators and stewards. The order of succession within each category shall be determined in accordance with the rank assigned by the operator.

Article 2.

(1) Within the periods specified in article 5 below, the aircraft commander:
(a) shall be in charge of the aircraft, the crew, the passengers, and the cargo;
(b) has the right and the duty to control and direct the crew and the passengers to the full extent necessary to ensure order and safety;
(c) has the right, for good reason, to disembark any number of the crew, or passengers at an intermediate stop;
(d) has disciplinary power over members of the crew within the scope of their duties; in case of necessity, of which he shall be sole judge, he

---

[1] ICAO DOC. 4006.

may assign temporarily any member of the crew to duties other than those for which he is engaged.

Article 3.

(1) The aircraft Commander shall have the right, without special authority:

a) to buy any items necessary for the completion of the trip;

b) to have any repairs made which are necessary to enable the aircraft to proceed promptly on its trip;

c) to make any arrangements and to undertake any expenditure which may be necessary for securing the safety of the passengers and crew and the preservation of the cargo;

d) to borrow the sums required for the accomplishment of the measures mentioned in paras. a), b) and c) of this article;

e) to engage, for the duration of the trip, in replacement of members of the crew who cease to be available for any reason, such personnel as is essential for the completion of the trip.

Article 4.

(1) The Commander may not, without special authority, sell the aircraft, or, by any contractual act, mortgage or subject it to any similar claim.

Article 5.

(1) The beginning and the end of the period during which the Commander maintains disciplinary control over the crew may be fixed by the operator. In any case he is entitled to exercise such control as soon as the crew embarks. At all stopping places, including the end of the trip, he continues to be so entitled at least until the fomalities of arrival are completed or until his command is taken over by another person.

(2) The powers of the Commander over the aircraft, the passengers and the cargo on board come into force as soon as the aircraft, with passengers and cargo, are handed over to him at the beginning of the trip. They expire at the end of the trip when the aircraft, the passengers and the cargo have been respectively handed over to the operator's representative or other qualified authority.

Article 6.

(1) In all countries and under all circumstances the Commander shall have the right of access to:

a) the Consul of the nationality of any person on board;

b) the Consul of the State in which the aircraft is registered;

c) the Consuls of the States of either party to a charter of the aircraft, and of consignors of cargo.

(2) After hearing the Commander, the Consuls may take any necessary means which are in accordance with the laws and consular regulations of their respective States.

(3) If the Commander first refers to the Consul of a State other than the State in which the aircraft is registered, the Commander shall notify

this action as soon as possible to the nearest Consul of the State in which the aircraft is registered.

Article 7.

(1) Births and deaths occurring on board the aircraft shall be recorded in the Journey log-book by the Commander, who shall issue extracts to the parties interested. He shall as soon as possible transmit certified extracts to the competent authority of the State in which the aircraft is registered and to that of the place of first landing, if so requested by the local authorities.

Article 8.

(1) The provisions of this Convention do not affect any international conventions or the laws or regulations of the Contracting States defining the conditions of qualifications required of an aircraft Commander.

Article 9.

(1) This Convention shall be applied in the case of an aircraft performing an international flight provided that such aircraft is registered in one of the Contracting States, or is operated by a national of any such State.

Article 10.

(1) This Convention does not apply to the Commander serving on board military, customs or police aircraft.

Article 11.

(1) This Convention is drawn up in the English, French and Spanish languages (each of which shall be of equal authenticity) in a single copy in each language. These three copies shall remain deposited in the archives of the International Civil Aviation Organization. One duly certified copy of this Convention shall be sent by the Secretary-General of the International Civil Aviation Organization to the Government of each signatory State.

Article 12.

(1) The Contracting States undertake to furnish the Secretary-General of the International Civil Aviation Organization with the laws and regulations enacted in pursuance of the provisions of this Convention as soon as they come into force. The Secretary-General shall send copies of these laws and regulations to all Contracting States.

Article 13.

(1) This Convention shall be ratified. Instruments of ratification shall be deposited in the archives of the International Civil Aviation Organization, which will notify the deposit to the Governments of each of the signatory States.

(2) As soon as this Convention shall have been ratified by two of the signatory States, it shall come into force as between them on the ninetieth day after the deposit of the second ratification. Thereafter, it shall come into force between the States which have already ratified and any other State which deposits its instrument of ratification on the ninetieth day after such deposit.

(3) The International Civil Aviation Organization shall notify to the Governments of each of the Contracting States the date on which this Convention comes into force and the date of deposit of each ratification.

(4) This Convention, as soon as it comes into effect, shall be registered with the United Nations.

Article 14.

(1) This Convention shall, after it has come into force, be open for adherence by any non-signatory State.

(2) Adherence shall be effected by a notification addressed to the International Civil Aviation Organization, which will inform the Governments of each of the Contracting States.

(3) Adherence shall take effect as from the ninetieth day after notification thereof to the International Civil Aviation Organization.

Article 15.

(1) Any Contracting State may denounce this Convention by notification addressed to the International Civil Aviation Organization, which will at once inform the Governments of each of the Contracting States.

(2) Denunciation shall take effect six months after the notification of denunciation, and shall operate only as regards the State which shall have denounced.

Article 16.

(1) Any Signatory or Contracting State may, at the time of signature or the deposit of ratification or of adherence, declare that the acceptance which it gives to this Convention does not apply to all or any of its Colonies, Protectorates, Mandated Territories or any territory subject to its sovereignty or authority, or any territory under its suzerainty or trusteeship.

(2) Any Contracting State may subsequently adhere separately on behalf of all or any of its Colonies, Protectorates, Mandated Territories or any territory subject to its sovereignty or authority or any territory under its suzerainty or trusteeship which has been excluded as aforesaid by its original declaration.

(3) Any Contracting State may denounce this Convention, in accordance with its provisions, separately or for all or any of its colonies, Protectorates, Mandated Territories or any territory subject to its sovereignty or authority, or any territory under its suzerainty or trusteeship.

Article 17.

(1) Upon deposit of ratification or adherence, or at any time thereafter any Contracting State may declare that any territory subject to its

sovereignty, protectorate, suzerainty, mandate, authority or trusteeship is to be regarded as a separate Contracting State for the purpose of this Convention.

(2) The International Civil Aviation Organization shall notify to the Governments of each of the Contracting States every declaration made in pursuance of paragraph (1) of this Article, and the date of such declaration.

(3) Any declaration made as aforesaid shall become effective, in relation to the territory concerned, on the ninetieth day after the date of such declaration.

(4) Any such declaration may be rescinded by notification thereof to the International Civil Aviation Organization, which shall notify such rescission to the Governments of each of the Contracting States, and the date thereof. Such rescissions shall become effective on the ninetieth day after the date of rescission.

Article 18.

Any difference between two or more Contracting States, relating to the interpretation or the application of this Convention, shall be settled, as provided by Chapter XVIII of the Convention on International Civil Aviation open for signature at Chicago on the 7th day of December, 1944.

This Convention shall remain open for signature until the   day of   One thousand nine hundred and

DONE at   the   day of   One thousand nine hundred and

IN WITNESS WHEREOF the undersigned plenipotentiaries, duly authorized, signed this Convention on behalf of their respective Governments.

— E N D —

ABBREVIATIONS

A.T.C.	Air Traffic Control
Archf. L.R.	Archiv für Luftrecht
B.W.	Burgerlijk Wetboek (Dutch Civil Code)
C.A.A.	Civil Aeronautics Administration
C.A.B.	Civil Aeronautics Board
C.I.N.A.	Commission Internationale de la Navigation Aérienne
C.I.T.E.J.A.	Comité International Technique d'Experts Juridiques Aériens
G.C.A.	Ground Controlled Approach
I.A.T.A.	International Air Transport Association
I.F.A.L.P.A.	International Federation of Airline Pilots Associations
I.F.R.	Instrument Flight Rules
I.L.A.	International Law Association
I.L.O.	International Labour Organization
J.A.L.	Journal of Air Law and Commerce
J.R.Ae.S.	Journal of the Royal Aeronautical Society
K.L.M.	Koninklijke Luchtvaart Maatschappij (K.L.M. Royal Dutch Airlines)
L.V.R.	Luchtverkeersreglement (Dutch Air Traffic Regulations)
M.v.A.	Memorie van Antwoord (Memorandum in Reply)
M.v.T.	Memorie van Toelichting (Explanatory Memorandum)
N.J.B.	Nederlands Juristenblad
N.J.	Nederlandsche Jurisprudentie
O.A.C.I.	Organisation de l'aviation civile internationale
(P.) I.C.A.O.	(Provisional) International Civil Aviation Organization
R.Ae.S.	Royal Aeronautical Society
R.D.A.	Revue de Droit Aérien
R.F.D.A.	Revue Française de Droit Aérien
R.D.I.L.C.	Revue de droit international et de législation comparée
R.G.A.	Revue générale de l'air
R.G.D.A.	Revue Générale de Droit Aérien
Riv. dir. aer.	Rivista di diritto aeronautico
R.T.L.	Regeling Toezicht Luchtvaart (Dutch Regulations for State Control of Air Navigation)
S.	Staatsblad (Dutch Statute Book)
U.K.	United Kingdom
V.F.R.	Visual Flight Rules
W.v.K.	Wetboek van Koophandel (Dutch Commercial Code)
W.v.S.	Wetboek van Strafrecht (Dutch Criminal Code)

BIBLIOGRAPHY

ARMOUR, MERRILL and HARLEY G. MOORHEAD Jr., Analysis of the Civil Aeronautics Boards Precedents in Safety Enforcement Cases. JAL 1950, p. 54.

BABINSKI, L., l'Aspect juridique de la notion du Commandant de l'aéronef. Riv. Dir. Aer. 1932, p. 412.

BAKER, MORRIS B., Airline Traffic and Operations. New York 1947.

BEAUMONT, K. M., Need for revision and amplification of the Warsaw Convention. JAL 1949, p. 395.

——, Report on Legal Position of the Aircraft Commander. IATA Bulletin 1933, no. 19, p. 18.

BERGIN, K. G., The Physiological Aspects of Air Safety. J.R.Ae.S. October 1949.

BOND, DOUGLAS D., The Love and Fear of Flying. New York 1952.

BRABAZON of TARA, Landing and Taking-off of aircraft in bad weather. Command paper 8147, London 1951.

BRATSCHI, HEINZ., Die Rechtsstellung des Luftfahrtpersonals. Bern 1951.

BUCHER, Pierre, Le Statut juridique du personnel navigant de l'aéronautique civile. Lausanne 1949.

BULLOCK, RAYMOND, Airline Piloting. Denver 1947.

CHARLIER, R. E., Le Commandant d'aéronef en droit privé. RGDA 1947, p. 20.

CHAUVEAU, P., Droit Aérien. Paris 1951.

CITEJA, Comptes rendus des Sessions,

——, Comptes rendus des Réunions de la 4ème Commission,

——, Rapport de M. Thieffry,

——, Rapports de M. Babinski,

——, Rapports de M. Garnault,

——, Other documents referred to in the text.

CLEVELAND, REGINALD M., Air Transport at War. New York 1946.

CLEVERINGA, R. P., Het Nieuwe Zeerecht. Zwolle 1946.

COOLING, R. D., Aircraft despatch and flight supervision. Aeronautics, Aug. 1950, p. 24.

COOPER, JOHN C., The Right to Fly. New York 1947.

——, The Legal Status of Aircraft. Princeton 1949 (stencil).

——, Air law – A Field for International Thinking. Transport and Communications Review, Oct.–Dec. 1951, p. 6.

COQUOZ, R., Le Droit privé international aérien. Paris 1938.

——, Les Perspectives d'avenir du Droit Privé International. RGDA 1938, p. 29.

CRAVEN, WESLEY F. and JAMES L. CATE, The Army Air Forces in World War II. Chicago 1948.

DANI LOVICS-SZONDY, Les infractions à la loi pénale commises à bord des aéronefs. Droit Aérien 1930, p. 402.

12

DAVIS, D. RUSSELL, Pilot Error, Some Laboratory Experiments. London 1948.

DÖRING, Das Arbeitsrecht des Bordpersonals der deutschen Luftfahrt-unternehmen. ArchfLR 1941.

——, Sozialversicherung. ArchfLR 1937.

DIJKSTRA, GERALD O. and LILIAN G. DIJKSTRA, The Business Law of Aviation. New York 1946.

EASTMAN, SAMUEL E., Liability of the ground control operator for negligence. JAL 1950, p. 170.

FIXEL, ROWLAND W., The Law of Aviation. Charlottesville 1948.

FREDERICK, JOHN H., Commercial Air Transportation. Chicago 1947.

FRESE, JÜRGEN, Fragen des internationalen Privatrechts der Luftfahrt unter besonderer Berücksichtigung einer Anwendungsmöglichkeit des Flaggenrechts. Köln 1940.

GIANNINI, AMEDEO, Lo Stato Giuridico della Gente dell'Aria. Roma 1937.

GOEDHUIS, D., Handboek voor het Luchtrecht. 's Gravenhage 1943.

——, La Convention de Varsovie du 12 October 1929. Leiden 1933.

——, National Air Legislations and the Warsaw Convention. 's Graven-hage 1937.

——, Air Law in the Making. 's Gravenhage 1938.

——, Idee en Belang in de Internationale Luchtvaart. 's Gravenhage 1947.

—— La Situation juridique du Commandant de l'aéronef. R.D.I.L.C. 1933, p. 134.

LE GOFF, M., Traité Theorique et Practique de Droit Aérien. Paris 1934.

——, La Loi du 25 Mars 1936 et le statut du Personnel navigant de l'aéronautique civile. R.G.D.A. 1936, p. 145.

——, The present state of airlaw. 's Gravenhage 1950.

GORDON, THOMAS, The Airline Pilot: A Survey of the critical requirements of his job and of pilot evaluation and selection procedures. CAA Division of Research, Report No. 73, Washington D.C. 1947.

GULDIMANN, Flugunfalluntersuchungen. Schweizer Aero Revue 1951 and 1952.

VAN HOUTTE, JEAN, La Responsabilité civile dans les transports aériens intérieurs et internationaux. Louvain 1940.

HEYT, EDWARD B., The human equation in aircraft accidents. CAB, Washington D.C.

HONIG, J. P., Overheidsaansprakelijkheid en Luchtvaart. NJB 1951, p. 767.

——, De positie van de Gezagvoerder van een luchtvaartuig. NJB 1951, p. 317.

HILBERT, Jurisdiction in High Seas Criminal Cases. JAL 1951, p. 427 and 1952, p. 25.

ICAO, Annexes to the Convention of Chicago,

——, Minutes and Documents Legal Committee,

——, Final Reports Division Meetings,

——, Other documents referred to in the text.

JAMES, J. W. G., Air Safety from the Pilots Point of View. J.R.Ae.S. October 1949.

DE JUGLART, M., Traité élémentaire de Droit Aérien. Paris 1952.

KNAUTH, A. W., The Aircraft Commander in International Law. JAL 1947.

KOFFKA-BODENSTEIN-KOFFKA, Luftverkehrsgesetz und Warschauer Abkommen.
KROELL, JOSEPH, Traité de Droit International Public Aérien. Paris 1934.
LATCHFORD, STEPHEN, Private International Airlaw. Department of State Bulletin 1945, p. 11.
——, Coordination of CITEJA with the new International Civil Aviation Organizations. Department of State Bulletin 1945, p. 310.
——, Private International Airlaw Developments. Department of State Bulletin 1946, p. 879.
LEMOINE, MAURICE, Traité de Droit Aérien. Paris 1947.
LONGHURST, JOHN, Jets and Airtraffic Control. The Aeroplane, Aug. 18 1950, p. 195.
LUPTON Jr., GEORGE W., Civil Aviation Law. Chicago 1935.
MAMORU MOCHIZUKI, Aviation Psychology. CAA Division of Research, Report No. 87, Washington D.C. 1949.
MANDL, VLADIMIR, La "Nationalité" des Aéronefs n'est qu'une dénomination erronée, Droit Aérien 1931, p. 161.
MANION, CLARENCE E., Law of the Air. Indianapolis 1950.
MASCHINO, MAURICE, La Condition juridique du Personnel Aérien. Paris 1930.
McFARLAND, ROSS A., Human Factors in Air Transport Design. New York 1946.
MEYER, ALEX, Freiheit der Luft als Rechtsproblem. Zürich 1944.
——, Crimes et délits à bord des aéronefs. RGDA 1946, p. 544.
MOSS, WILLIAM W., How a Veteran Pilot looks at Safety. American Aviation, Febr. 18 1952, p. 21.
NEWTON, J. A., The Human Factor in Aircraft Accidents. J.R.Ae.S. February 1951, p. 110.
NICHOLSON, JOSEPH L., Air Transportation Management. New York 1951.
NIEMEYER, Crimes et délits à bord des aéronefs. RDA 1929, p. 285.
OGBURN, WILLIAM F., The Social Effects of Aviation. Cambridge Mass. 1946.
OPPIKOFER, HANS, Internationale Luftprivatrechtskonferenz. ArchfLR 1933, p. 211.
PARKER VAN ZANDT, J., Civil Aviation and Peace. Washington D.C. 1944.
——, The Geography of World Air Transport. Washington D.C. 1944.
PÉPIN, E., Le Droit Aérien. Paris 1947.
——, ICAO and other agencies dealing with air regulation. JAL 1952, p. 152.
PHOLIEN, Des Crimes et Délits à bord d'aéronefs en vol. RDA 1929, p. 289.
PLESMAN, A., Thirty years of civil aviation. IATA Bulletin No. 11, 1950.
PUFFER, CLAUDE E., Air Transportation. Philadelphia 1941.
RHYNE, CHARLES S., Aviation Accident Law. Washington D.C. 1947.
RICHTER, LUTZ, Bemerkungen zum Arbeitsrecht des Luftverkehrspersonals. ArchfLR 1935.
——, Die Tarifvertraglichen Arbeitsbedingungen des Italienischen Flugpersonals. ArchfLR 1932.
RIESE, OTTO, Luftrecht. Stuttgart 1949.
——, Die 6. Jahresversammlung der CITEJA, Paris 1931. ArchfLR 1932.
RIJKS, K., Het Verdrag van Genève, betreffende internationale erkenning van rechten op luchtvaartuigen. Diss. Leiden 1952.

SANDIFORD, ROBERTO, Lo Stato giuridico del Commandante di aeromobile. Roma 1934.

SAUVEPLANNE, J. G., Luchtvaartverzekering. Diss. Leiden 1949.

SAVOIA, La Figura giuridica del Commandante di aeromobile. Riv. Dir. Aer. 1929, p. 193.

SHAWCROSS, C. N. and K. M. BEAUMONT, Air Law. London 1950, with supplement 1952.

SIMPSON, Use of aircraft accident investigation information in actions for damage. JAL 1950, p. 283.

SPEAS, R. DIXON, Airline Operation. Washington D.C. 1948.

STEWART, MAJOR OLIVER, Airpower and the expanding Community. London 1944.

SURVIVAL IN THE AIR AGE, a Report by the Presidents Air Policy Commission. Washington 1948.

SWEENEY, Safety regulations and accident investigation. JAL 1950, p. 161 and p. 269.

SYMONDS, Sir CHARLES P. and DENIS J. WILLIAMS, Psychological Disorders in Flying Personnel of the Royal Air Force, investigated during the war 1939–1945. London 1947.

TWEEDIE e.a., Air Accident Investigation and human failure, lectures for the Royal Swedish Academy of Engineering Science. Stockholm 1951.

VEAL, J., Some British Views on Flight Safety. Third Annual Safety Seminar, Flight Safety Foundation, October 1950.

VERSCHOOR, I. H. PH., Het Verdrag van Brussel van 1938 betreffende hulp en berging van of door luchtvaartuigen op zee. 's Gravenhage 1943.

DE VISSCHER, F., Le Règlement des compétences pénales en Droit Aérien. RGDA 1937, p. 329.

——, Les conflits de lois en matière de droit aérien. Recueil des Cours de l'Academie de droit international à la Haye 1934, T. II p. 279.

VOLKMAN, Crimes et délits à bord des aéronefs en droit international. Droit Aérien 1931, p. 26.

WAGNER, WIENCZYSLAW, Les Libertés de l'Air. Paris 1948.

WILHELM, De la Situation juridique des aéronautes en droit international. Journal du droit International Privé 1891, p. 440.

WILSON, G. LLOYD and LESLIE A. BRYAN, Air Transportation. New York 1949.

WOLFE, THOMAS, Air Transportation — Traffic and Management. New York 1950.

WOLFSBERGEN, Processuele Curiosa bij de Raad voor de Scheepvaart. NJB 1938, p. 82.

WRIGHT, T. P., Aviation's Place in Civilization. R.Ae.S. repr. No. 101, 1945.

——, Research and Development to promote Safety in Aviation. Society of Automotive Engineers Quarterly Transactions, April 1951, Vol. V. No. 2, p. 173.

WÜSTENDORFER, Bericht des Seeluftrechtlichen Ausschusses des Deutschen Nautischen Vereins über die Rechtsgestaltung des überseeischen Luftverkehrs. ArchfLR 1931.

ZOLLMAN, CARL, Law of the Air. Milwaukee 1927.

ZONDAG, C. H. G. M., Neutraliteit in de Lucht. 's Gravenhage 1940.

ZWENG, CHARLES A., Airline Transport Pilot Rating. Hollywood 1947.

INDEX

SAMENVATTING

Het gecompliceerde mechanisme van de moderne verkeersluchtvaart kan slechts functioneren door de toewijding van talloos velen.

Onder hen, die in deze gezamenlijke arbeid een taak vervullen, neemt de gezagvoerder van een luchtvaartuig echter een bijzondere plaats in.

Hij ziet zich aan het hoofd geplaatst van een weliswaar kleine, doch betrekkelijk geïsoleerde gemeenschap, die zich in een kort tijdsbestek onder verschillende jurisdicties kan verplaatsen.

In verschillende opzichten neemt de gezagvoerder in de luchtvaart een sleutelpositie in, aangezien de veiligheid, de economie en de regelmaat van de vlucht vaak in belangrijke mate afhangen van zijn inzicht en bekwaamheid.

Anders dan de scheepskapitein, die een zowel op het gecodificeerde recht als op de traditie berustende rechtspositie inneemt, ontbeert de gezagvoerder van een luchtvaartuig echter een nauwkeurig omlijnde status.

Aangezien het luchtrecht, als wellicht geen ander onderdeel van het recht, nog in beweging is, is het haast onvermijdelijk dat een studie op dit gebied het karakter van een momentopname heeft.

Als object hiervoor is in het eerste deel van deze studie gekozen de gezagvoerder van een luchtvaartuig in het huidige Nederlandse recht.

In het publiekrecht is de gezagvoerder van een luchtvaartuig aan een vèrgaande reglementering onderworpen, waarbij hem bij de voorbereiding van een vlucht, bij de uitvoering ervan en na afloop talloze verplichtingen worden opgelegd. Gezien het internationale karakter van de luchtvaart is unificatie van voorschriften van groot belang. De hiertoe door de ICAO [1] in de Annexen van het Verdrag van Chicago uitgevaardigde "Standards and

---

[1] International Civil Aviation Organization.

Recommended Practices" vormen echter geen dwingend recht en hoewel zij in het algemeen, behoudens enkele afwijkingen, worden toegepast, missen zij vaak nog wettelijke kracht. De Nederlandse luchtvaartwetgeving, die slechts boven Nederlands grondgebied geldt, bevindt zich in een onoverzichtelijk overgangsstadium, aangezien aanpassing aan de normen van de ICAO nog niet heeft plaats gevonden.

In enkele gevallen, zoals bij het bijhouden van de boorddocumenten en het uitvoeren van sanitaire bepalingen wordt de gezagvoerder ingeschakeld als orgaan van de overheid. Zijn taak in dit opzicht gaat echter aanmerkelijk minder ver dan die van de scheepskapitein, die bijv. politionele bevoegdheid heeft, een taak bij het registreren van geboorte en overlijden etc. Wat de lacunes in de rechtspositie van de gezagvoerder betreft, werd gewezen op het ontbreken van een algemene wettelijke bevoegdheid tot het uitoefenen van gezag over de inzittenden. Evenmin bleek zijn bevoegdheid om als vertegenwoordiger van de ondernemer op te treden wettelijk geregeld te zijn.

Bij een beschouwing van het vehicle, waarin de gezagvoerder zijn dagelijkse werkzaamheden verricht, het verkeersvliegtuig, bleek ons, hoe moeilijk het is een exacte scheiding te maken tussen burger- en militaire luchtvaartuigen. Voorts bleek, dat het luchtvaartuig een nationaliteit bezit, doch dat aangaande de aan boord toepasselijke wet nog geenszins eenstemmigheid heerst.

Volgens het huidige Nederlandse recht is een van de bemanningsleden de gezagvoerder, doch dit behoeft niet een vliegtuigbestuurder te zijn. Opgemerkt werd, dat bij de uitgifte van vliegbewijzen slechts op physieke conditie, kennis en vaardigheid van de candidaat, doch niet op karaktereigenschappen gelet wordt. Wij kwamen tot de conclusie, dat het aanbeveling verdient de Nederlandse wetgeving op dit punt te wijzigen.

De onafhankelijkheid van de gezagvoerder is aanmerkelijk verminderd door de ontwikkeling van de telecommunicatiemiddelen, welke tijdens de vlucht een voortdurende verbinding met instanties op de grond mogelijk maken. De bevoegdheid om tijdens de vlucht instructies aan de gezagvoerder te geven van "Air Traffic Control", "Operational Control" en de havenmeester werd aan een beschouwing onderworpen. De vraag of de gezagvoerder gebonden is instructies van de grond op te volgen, is vooral van belang, indien hij zich voor de beslissing gesteld ziet bij ongunstige weersomstandigheden een landing al dan niet uit te voeren.

De taak van de gezagvoerder bij hulp en berging is omschreven in het Verdrag van Brussel, dat echter niet is geratificeerd en in Annex 12 van het Verdrag van Chicago. De voor- en nadelen van beide regelingen werden besproken en wij kwamen tot de gevolgtrekking, dat het wenselijk is de meest bruikbare elementen samen te voegen.

De gezagvoerder kan getroffen worden door sancties van de strafrechter, alsmede door disciplinaire maatregelen van de Raad voor de Luchtvaart. Laatstgenoemd lichaam kan ook de bevoegdheid tot het uitoefenen van een functie aan boord van een luchtvaartuig intrekken, indien het van oordeel is, dat de betrokkene hiertoe de geschiktheid mist. Hoewel de samenstelling en de procedure van de Raad een aantal waarborgen geven voor een juiste rechtsbedeling, lijkt het gewenst de mogelijkheid van hoger beroep in te voeren, welke thans ontbreekt.

De aansprakelijkheid van de vervoerder en de ondernemer is geregeld in het Verdrag van Warschau en het Verdrag van Rome, doch deze Verdragen regelen of in het geheel niet, of slechts gedeeltelijk de aansprakelijkheid van de gezagvoerder, welke dus veelal door het gemene recht bepaald wordt. Het bijzondere karakter van de luchtvaart maakt o.i. een bijzondere en volledige regeling gewenst. De relatief grote frequentie, waarmede luchtvaartongevallen aan een fout van de bestuurder worden toegeschreven, leidde ons tot een beschouwing van de hierbij toe te passen normen en het verband met de aansprakelijkheid van vervoerder, ondernemer en gezagvoerder zelf.

Tenslotte bleek ook de positie van de gezagvoerder als werknemer door het gemene recht beheerst te worden. Hoewel de werkkring van de gezagvoerder een aantal bijzondere aspecten heeft, bestaat er o.i. echter niet de behoefte aan een bijzondere wettelijke regeling terzake.

Uit het voorgaande kan naar onze mening wel de conclusie getrokken worden, dat de status van de gezagvoerder van een luchtvaartuig vele leemtes en onzekerheden bevat en dat zijn positie niet alleen in letterlijke zin vaak ,,zwevende'' is.

SAMENVATTING

Indien men een luchtvaartuig tijdens de vlucht met de zich erin bevindende personen en zaken als een tijdelijk uit de samenleving gelichte kleine gemeenschap beschouwt, is het duidelijk dat deze gemeenschap een wettelijke regeling behoeft. Het internationale karakter van de luchtvaart maakt het gewenst een dergelijke regeling internationaal geldend te maken.

Het tweede deel van onze studie werd gewijd aan de pogingen om tot een Conventie aangaande de rechtspositie van de gezagvoerder te geraken. Reeds gedurende meer dan 25 jaar heeft dit onderwerp een punt van studie van het CITEJA [1] en thans van de ICAO gevormd en talloze ontwerpen zijn in deze periode verschenen.

Het ontbreken van precedenten in het luchtrecht heeft ertoe geleid, dat de positie van de gezagvoerder van een luchtvaartuig vergeleken is met die van de scheepskapitein, doch ook wel met die van een schipper in de binnenscheepvaart, chauffeur van een autobus, of conducteur van een trein. In al deze gevallen gaat het om vervoermiddelen, welke het vermogen bezitten zich over de landsgrenzen te bewegen, waarbij zij vaak grote afstanden afleggen en een betrekkelijk geïsoleerde groep personen vervoeren.

Het wil echter voorkomen, dat er bij deze vergelijkingen meer punten van verschil dan van overeenkomst bestaan.

Wat de vergelijking met binnenschepen, autobussen en treinen betreft, zij er op gewezen, dat het isolement van de zich hierin bevindende personen aanzienlijk minder groot is dan bij een luchtvaartuig; deze vervoermiddelen bevinden zich altijd op het gebied van andere Staten, nimmer op statenloos gebied; zij bewegen zich langs vaste tracé's; de betrokken Staat kan te allen tijde zijn gezag feitelijk doen gelden en ingrijpen in de rechtssfeer aan boord.

Wat betreft de vergelijking met de scheepskapitein is een zekere analogie onmiskenbaar. Toch zijn er verschillen van principiële en graduele aard: de duur van een luchtreis is over het algemeen aanzienlijk korter dan een zeereis; de bemanning van een luchtvaartuig is aanmerkelijk kleiner dan die van een zeeschip, het aantal passagiers meestal geringer; de beperkte bewegingsvrijheid

---

[1] Comité International Technique d'Experts Juridiques Aériens.

in een luchtvaartuig geeft de samenleving een geheel ander karakter dan die aan boord van een schip; de in de zeevaart belangrijke traditie ontbreekt in de luchtvaart. Om al deze redenen komt het gewenst voor, bij het formuleren van een statuut voor de gezagvoerder van een luchtvaartuig niet uit te gaan van schijnbare overeenkomsten met andere vervoermiddelen, doch het luchtvaartuig als een vehicle *sui generis* te beschouwen en de gezagvoerder een rechtspositie te geven in overeenstemming met de bijzonderheden van de luchtvaart.

Het ontwerp van de ICAO voor een Conventie tot regeling van de rechtspositie van de gezagvoerder van een luchtvaartuig schijnt aan dit postulaat te voldoen en komt ons in het algemeen aanvaardbaar voor. Het lijkt echter wenselijk het Verdrag ook van toepassing te doen zijn op nationaal verkeer.

Behoudens een aantal wijzigingen en aanvullingen van minder principiële aard, dient het ontwerp o.i. in twee opzichten te worden aangevuld. In de eerste plaats komt het gewenst voor in het Verdrag de verplichting van de gezagvoerder vast te leggen, om te zorgen voor een veilige uitvoering van de vlucht. Voorts dient de aansprakelijkheid van de gezagvoerder beperkt te worden tot het geval er sprake is van opzet of grove schuld.

Onze conclusie luidt, dat het in hoge mate wenselijk is, dat thans spoedig een Verdrag tot stand zal komen, dat de gezagvoerder een status verleent, in overeenstemming met de belangrijke en verantwoordelijke taak welke deze in de verkeersluchtvaart vervult.